Transportation Planning and Management in Developing Economies

Transportation planning and management addresses the complex mobility challenges faced by both large metropolitan cities and rural areas. This updated edition delves deeper into critical aspects of urban and rural mass transportation planning and management, offering a comprehensive and balanced perspective. Each chapter has been enriched with updated time-series data and an expanded collection of practical examples to illustrate key concepts effectively. The theoretical sections have been significantly enhanced to include detailed discussions on transit operating elements, the planning processes for mass transit systems, and advanced management methods for mass transportation, with a particular focus on the integration of emerging technologies.

- Contains exclusive material on transport planning in developing economies
- Discusses both planning and management aspects
- Explores data from urban and rural public transport
- Examines resource needs and economics of public transport
- Includes examples and case studies from BRIC countries

This book is aimed at senior undergraduate and graduate students in transportation engineering and for transport planners.

Transportation Planning and Management in Developing Economies

Focus on Mass Public Transportation

Second Edition

Ashish Verma and T. V. Ramanayya

CRC Press
Taylor & Francis Group
Boca Raton London New York

CRC Press is an imprint of the
Taylor & Francis Group, an **informa** business

Designed cover image: Getty Images

Second edition published 2027
by CRC Press
2385 NW Executive Center Drive, Suite 320, Boca Raton FL 33431

and by CRC Press
4 Park Square, Milton Park, Abingdon, Oxon, OX14 4RN

CRC Press is an imprint of Taylor & Francis Group, LLC

First edition published by CRC Press, 2015

Library of Congress Cataloging-in-Publication Data
Names: Verma, Ashish author | Ramanayya, T. V. author
Title: Transportation planning and management in developing economies : focus on mass public transportation / Ashish Verma and T.V. Ramanayya.
Other titles: Public transport planning and management in developing countries
Description: Second edition. | Boca Raton, FL : CRC Press, 2026. | Includes bibliographical references and index.
Identifiers: LCCN 2025049244 (print) | LCCN 2025049245 (ebook) | ISBN 9781032661995 hardback | ISBN 9781041241270 paperback | ISBN 9781003740247 ebook
Subjects: LCSH: Transportation—Developing countries—Planning
Classification: LCC HE148.5 .V47 2026 (print) | LCC HE148.5 (ebook) | DDC 388.4068/4—dc23/eng/20260423
LC record available at https://lccn.loc.gov/2025049244
LC ebook record available at https://lccn.loc.gov/2025049245

ISBN: 978-1-032-66199-5 (hbk)
ISBN: 978-1-041-24127-0 (pbk)
ISBN: 978-1-003-74024-7 (ebk)

DOI: 10.1201/9781003740247

Typeset in Times
by Apex CoVantage, LLC

Dedicated
to
Lord Ganesha

Contents

About the Authors xi
Preface to the Second Edition xiii
Preface to the First Edition xv
Acknowledgments xvii

Chapter 1 Introduction 1

1.1 General 1
1.2 Demographic Changes of India Since Independence and Migration to Urban Areas 2
1.3 Urbanization 3
1.4 Growth of Urban Settlements 6
1.5 Urban Development in India 8
1.6 Migration to Urban Areas 8
1.7 Gross Domestic Product 10
1.8 Urban Transportation Status in Developing Countries 10
Summary 13

Chapter 2 Different Public Transport Modes, Their Characteristics, and Operating Elements 15

2.1 Modes of Transport 15
2.2 Transport Systems Definitions and Classifications 15
2.3 Transit Modes' Characteristics 24
2.4 Public Transport Modes and Classification of Mass Transit Modes 26
2.5 Operating Elements 39
2.6 Status of Public Transport with Respect to Coverage, Fleet Strength, and Utilization in Urban Areas 42
2.7 The Need for an Integrated Transport Infrastructure Approach 43
2.8 Comparative Analyses 44
2.9 A Look Ahead in Public Transportation 47
Summary 48

Chapter 3 Current Scenario of Public Mass Transport (PMT) in Developing Countries 49

3.1 Current Perception and Usage of PMT in the Developed and Developing Nations 49
3.2 Case Study 50

3.3 Planning and Operational Strategies Followed in India and in Other Developing Countries ... 57
3.4 Issues of Coordination with Respect to First-Mile and Last-Mile Connectivity and LOSs ... 62
3.5 Funding Pattern ... 71
Summary ... 76

Chapter 4 Public Mass Transport Planning in Urban Areas ... 78

4.1 Urban Travel Demand Estimation for Different Categories of Urban Areas for the Next Two Decades ... 78
4.2 Current Transport Infrastructure and Gap Identification with Respect to Short-, Medium-, and Long-Term Estimated Demand Patterns ... 113
4.3 Routing and Scheduling of PMT Services ... 116
4.4 Strategies Exhibited by General Public (Different Sections of the Public) with Regard to Choice of Mode ... 122
4.5 LOS Concept in Planning Urban Mobility ... 123
4.6 Stakeholder Consultation and Preferences with Respect to First-Mile and Last-Mile Connectivity ... 124
4.7 Feasibility for Higher-Order Modes Like BRT and Metro Rail ... 129
Summary ... 132

Chapter 5 Planning and Management of Urban Transport ... 133

5.1 Dedicated Organization for Continuous Urban Transport Management ... 133
5.2 Data Collection, Retrieval, and Analysis for Effective Management ... 156
5.3 Role of Private Sector in Public Transport Management ... 157
5.4 Assessment of User Satisfaction ... 167
5.5 The Global Trend ... 167
Summary ... 168
Appendix ... 169
Respondent Category ... 169

Chapter 6 Special Focus on Public Mass Transport Planning in Rural Areas ... 189

6.1 Rural Transport Scenario Changes and Their Impact on Economic Development and Demand for Improved Transport ... 189
6.2 PMGSY (Dedicated Rural Transport Development) ... 189
6.3 The Need for Network Planning in a Region ... 193
6.4 Travel Demand Estimation in a Region, Say District ... 203

6.5 Stakeholder Consultation and Preferences to Optimize PMT Routes ... 205
6.6 Impact of Transport Infrastructure in the Development of Rural India—A Case Study ... 207
Summary ... 213

Chapter 7 Economics of Public Mass Transport ... 214

7.1 Energy Efficiency of PMT with Respect to Percentage Contribution of PMT ... 214
7.2 Costing of PMT Services ... 218
7.3 Manpower Requirements in Different Categories, Such As Operations and Management in Both Rural and Urban Areas ... 219
7.4 Estimation of Social, Economic, and Environmental Benefits Associated with Improved PMT ... 224
Summary ... 233

Chapter 8 Resource Needs and Organizational Structure for Public Mass Transport at State Level ... 234

8.1 Current Financial Performance of PMT Organizations in India ... 234
8.2 Review of the Existing Funding Pattern ... 239
8.3 Estimation of Resource Requirements at State and Central Level for the Next Two Decades ... 239
8.4 Identification of Innovative Strategies to Generate Revenue in a Sustainable Manner ... 241
8.5 The Need for Dedicated Funding as a Percentage of State Revenue and Creation of State-Level Organization for Optimum Usage of These Resources ... 242
Summary ... 243

References and Bibliography ... 245

Index ... 251

About the Authors

Ashish Verma is a PhD from IIT Bombay and currently serving as Professor of Transportation Systems Engineering in the Department of Civil Engineering at the Indian Institute of Science (IISc), Bengaluru, India. Furthermore, he was Visiting Professor at ITMO University, Saint Petersburg, Russia, during 2016 and Visiting Fellow at Queensland University of Technology (QUT), Brisbane, Australia, during 2019. Before joining IISc, he had served in IIT Guwahati and the Mumbai Metropolitan Region Development Authority (MMRDA). His research interests include sustainable transportation planning and policy, integrated public transport planning and management, non-motorized transport (NMT), transport and livability/quality of life (QoL), transport and climate change, modeling and optimization of transportation systems, travel behavior analysis and modeling, pedestrian and crowd flow modeling, driver behavior and road safety, intelligent transportation systems (ITS), and traffic control and management. He has authored more than 260 research publications, including more than 100 journal publications, in the area of transportation systems engineering. He also has a patent and a software copyright to his credit. He is the founding editor-in-chief (EiC) of the new journal *Sustainable Transport and Livability* published by Taylor & Francis (T&F). He is the former editor of journals of several internationally recognized publishing houses. He is presently serving as a steering and scientific committee member of the World Conference on Transport Research Society (WCTRS) based in the University of Leeds, UK. He is also currently serving as Hony. Secretary of the Institute of Urban Transport (India).

T. V. Ramanayya holds a degree in civil engineering, a master's in transportation engineering, and a PhD in transportation engineering, all from the Regional Engineering College (presently known as the National Institute of Technology) Warangal, Andhra Pradesh. He has more than four decades of experience (7 years in REC Warangal and 32 years at the Indian Institute of Management (IIM) Bangalore (IIMB)) in teaching, training, research, and consultancy in the areas of transportation and urban infrastructure. He has guided a number of students for master's degree and PhD thesis. Some of his project assignments include holistic organizational transformation strategies (HOTS) for the Andhra Pradesh State Road Transport Corporation (APSRTC), sponsored by the government of Andhra Pradesh; development of strategic policymaking on bus passenger transport in India; design of internal control systems for the Karnataka State Road Transport Corporation (KSRTC); rural transport infrastructure in entire South India; assessment of social and economic impact of rural roads component, sponsored by the government of Andhra Pradesh. He was involved in developing two new courses and taught them to master's level students at IIMB, which include planning and management of urban infrastructure and computer applications, including simulation programs for urban traffic and transportation.

Preface to the Second Edition

Since the release of the first edition, there has been a noticeable surge in both interest and action in the public transport sector, particularly across cities in India and other developing nations. Rapid urbanization, the push for sustainability, the post-pandemic recovery, and the emergence of new technologies have all played a role in transforming how public transport is planned, implemented, and managed.

This second edition has been developed in response to these evolving challenges and opportunities. It reflects our ongoing commitment to provide a comprehensive, practical, and context-specific resource for students, researchers, practitioners, and policymakers engaged in the public transport sector.

Key enhancements in this edition include:

- Updated data and examples to reflect the latest developments in Indian and international contexts
- New sections on emerging technologies, including electric and autonomous public transport options
- Expanded discussion on post-COVID travel behavior and its implications for transport planning
- Inclusion of recent government initiatives and policy frameworks supporting public transport
- Revisions to several chapters for improved clarity and alignment with current academic curricula

We are grateful to the many readers, reviewers, and students who provided valuable feedback on the first edition. Their insights have helped shape this revised edition and reinforce its relevance to real-world issues in public transport planning and management.

Preface to the First Edition

In the early stages of growth of any city, the vehicular trips are all road-based and are confined to modes like cycles, personalized cars, two-wheelers, and intermediate public transport (IPT) modes like cycle rickshaws, tongas, taxis, three-wheelers, and tempos. As the city's population and size grow further, commuter trips tend to get concentrated on particular sections and routes and call for a larger transport unit like a mini-bus or a standard bus, which forms part of the public transport system. In larger cities, public transport systems play an increasingly important role. In cities like Kolkata, Mumbai, Delhi, and Chennai, suburban rail services carry sizeable volumes of commuter trips. In Kolkata, tramways also play a noticeable role besides the metro rail operation. With the increase in city complexities and advancement in technology, new systems like automated guided transit (AGT) will soon be visible in the metropolitan cities of India.

While making the selection of a particular mode of mass transport for any Indian city, the criteria should essentially be the volume and pattern of travel demand. But, besides this, the systems should also be judged by their suitability for Indian conditions; effectiveness and efficiency in dependably performing their designated role(s); flexibility of operation, with wide applicability; capital requirement; energy sources and consumption; environmental impact, from the point of view of pollution and noise; aesthetics, including image; available technology and indigenous capability; proven or nonproven technology; and future potential (especially in technological advancements).

This book attempts to cover various aspects of public transport planning and management, particularly in the context of India and other developing countries, which have much different transportation issues and requirements as compared to developed countries. Many of the developing countries are attempting to expand and upgrade their public transport systems to create sustainable solutions for their cities and countries, though within the constraints of poor land use and urban planning, haphazard growth of settlements, high population and population density, growing income and travel demand, etc. Therefore, it is quite timely to come up with this book that explains the principles of public transport planning and management that are relevant and suitable for developing countries' situations. The book covers both urban and rural public transport planning and management, as well as aspects of economics and finance related to public transport.

The highlight of the book is the use of rich data from India and developing countries and various examples from the research works of the authors to explain various aspects of public transport planning and management. In the opinion of the authors, this book will be a good text/reference material for courses on public transport offered in various universities at postgraduate and undergraduate levels.

Acknowledgments

The authors express their sincere thanks to Dr. Furqan A. Bhat for helping the authors in compiling references relevant for the book writing and also for typing and preparing book chapters. Due acknowledgment also to Dr. Gagandeep Singh from CRC Press whose continuous push and follow-ups helped the authors to complete this second edition of the book in a reasonable time. The authors are grateful to their respective families for their constant support, encouragement, and help in completing this book and sparing us the time to write this book.

1 Introduction

1.1 GENERAL

India is a country located in South Asia that is home to the second-largest population in the world (behind China), estimated at 1.3 billion as of 2021. India is the largest democracy in the world, its official form of government being a federal constitutional republic. It also has the seventh largest geographical area in the world at 3,287,263 sq. km. India is a big country; it accounts for about 2.4% of the total surface area of the world. India is nearly 20 times as large as Great Britain. Many of the Indian states are larger than several countries of the world. It became an independent nation on August 15, 1947 by gaining freedom from the British Empire by a peaceful resistance led by Mahatma Gandhi. Currently, India is comprised of 28 states, 8 union territories, and 806 districts. In terms of religion, Hinduism is followed by 72.03% of the population of the country. As for language, Hindi is the most spoken language. In addition, there are 16 languages declared as official languages for the purpose of communication within India. There is no other country in the world with such diversity. Thus, India may be considered as a continent rather than a country. Figure 1.1 indicates the Indian subcontinent with administrative boundaries of different states. India is predominantly rural, but it also has many densely populated metropolitan areas. Most of the land is inhabited except for the higher regions of the Himalayan mountains. Since the time of the Indus Valley Civilization, the key factor that settlers have considered in deciding where to live in India has been the ability to have continuous cultivation. This is mainly determined by water availability and soil fertility.

The world is increasingly looking at the developing nations as potential markets and is eagerly waiting to see how the nations perform in the coming years. An efficient transportation system is essential for a country's development. The contribution of gross domestic product (GDP) per million urban population is much higher than per million population of rural areas. It is further observed that the urban areas contribute to a large proportion of GDP of a country. The world today is increasingly urbanized. The challenge of a growing population and increasing vehicle ownership and usage casts a worrying shadow over projected emission of global greenhouse gases (GHGs). It is roughly estimated that half the global population lives in cities. By 2030, it is projected that developing nations would have more vehicles than developed nations. Developing economies have been a prominent player in the game of rapid urbanization and its consequent irrevocable change in climate. Booming economy, car manufacturers' inflow to India, aspirations to own a car, increasing travel distances, comfort and safety, governments encouraging policies (open car market, easy loan schemes), etc. are a few reasons for increasing motorization at a rapid rate. In terms of an increase in motorization rates, India has registered a rapid growth in the number of two-wheelers between 1997 and 2022. Two-wheelers contribute to 73% of total vehicle population. The total number of motor vehicles (MVs) in India

DOI: 10.1201/ 9781003740247-1

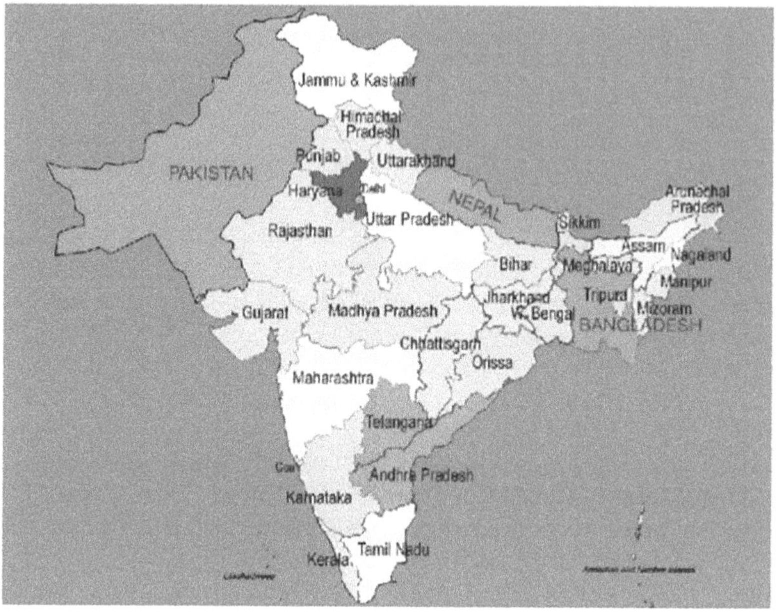

FIGURE 1.1 Map of India Indicating Different States

Source: www.mapsofindia.com

has increased from 55 million in 2000 to approximately 354 million in 2022. Motorized two-wheelers (motorcycles, scooters, mopeds) account for over 70% of the total registered fleet that will clearly impact carbon monoxide (CO) and hydrocarbon (HC) emissions. However, developing nations still possess the capacity for sustainable future in contrast to developed nations. Public transport and NMT still forms the major mode of travel. Unfortunately, the quality of these modes is often quite poor with regard to level of service (L/S), security, comfort, and convenience. The sum effect of inadequate public transport and difficult conditions for walking and cycling implies that most developing-city citizens will move to private motorized vehicles as soon as it is economically viable for them to do so. The basic function of a transport system is to provide mobility, flexibility to people, and accessibility to places. A sustainable transport system must offer mobility and accessibility to all urban residents in a secure and eco-friendly mode of transport. Fundamentally more efficient and sustainable transport systems in developing nations should emphasize preservation, improvement of existing modes, encouraging modal shift toward sustainable modes, and implementing proper planning measures.

1.2 DEMOGRAPHIC CHANGES OF INDIA SINCE INDEPENDENCE AND MIGRATION TO URBAN AREAS

The population of India as per 2011 census is above 1.2 billion and is estimated to be more than 1.3 billion in 2021. This population is distributed in around 580,000 rural

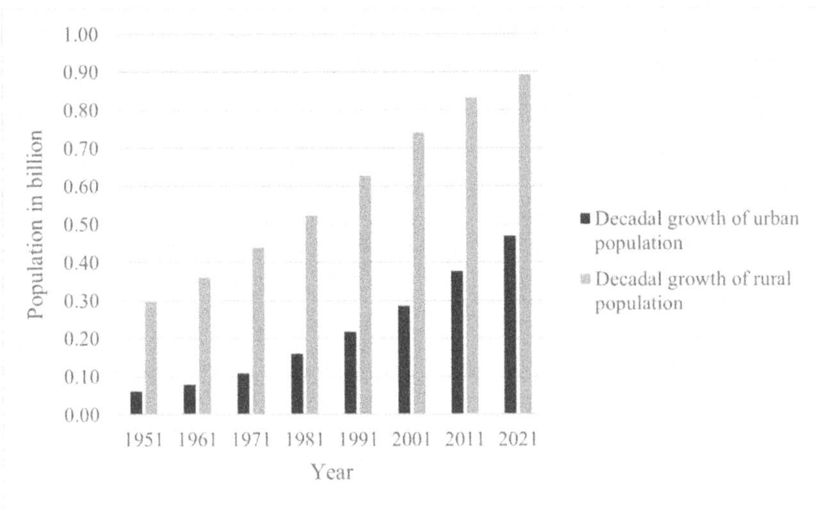

FIGURE 1.2 Decadal Growth of Urban and Rural Population

Source: Census of India 1951 to 2011, Population projections for Indian States (2011–2036), National Commission on Population Ministry of Health and Family Welfare

settlements and around 4500 urban centers of different sizes. The country is experiencing rapid growth in population. The growth of population since 1951 is presented in Table 1.1. The population of the country as per 2011 census has crossed 1.2 billion, compared to 0.361 billion during 1951. Figure 1.2 provides the pictorial presentation of urban and rural population from 1951 to 2011. The compounded annual growth rate (CAGR) during the past seven decades works out to be 2.24%. However, there are variations in the growth rates of urban and rural populations.

The CAGR of rural areas during the past seven decades is about 1.83%. The CAGR of urban areas during the past seven decades is about 3.49%.

It may be noted from Table 1.1 that urban population is increasing at a faster rate compared to rural population. It is further observed as presented in Table 1.1 that during the decade, that is, 2011–21, the net addition of the urban population (92,238,000) has exceeded the net addition of the rural population (59,913,000) by 32,325,000. This trend is likely to continue in the future. Demographers estimate that the percentage of people living in urban areas is likely to reach 50% of the total population by the year 2050.

1.3 URBANIZATION

An urban area is an area with an increased density of human-created structures in comparison to the areas surrounding it. Urban areas may be cities, towns, or conurbations, but the term is not commonly extended to rural settlements such as villages and hamlets. The definition of urban areas varies from nation to nation and, in India, an urban area is defined as towns (place with municipal corporation, municipal area

TABLE 1.1
Growth of Population Since 1951

Year	Total Population (in Billions)	Decadal % Growth of Total Population	Total Rural Population (in Billions)	% of Rural Population	Decadal Growth of Rural Population	Total Urban Population (in Billions)	% of Urban Population	Decadal Growth of Urban Population
1951	0.36	13.31	0.30	82.71	0.024	0.06	17.17	0.018
1961	0.44	21.64	0.36	82.03	0.062	0.08	17.97	0.016
1971	0.55	24.80	0.44	80.09	0.079	0.11	19.91	0.030
1981	0.68	24.66	0.52	76.66	0.085	0.16	23.34	0.050
1991	0.85	23.87	0.63	74.28	0.105	0.22	25.72	0.058
2001	1.03	21.54	0.74	72.17	0.114	0.29	27.83	0.069
2011	1.21	17.64	0.83	68.84	0.091	0.38	31.16	0.091
2021*	1.40	12.40	0.89	65.66	0.058	0.47	34.71	0.089

Source: Census of India for different years,

* Population projections for Indian States (2011–2036), National Commission on Population Ministry of Health and Family Welfare

committee, town committee, notified area committee, or cantonment board), all places having 5000 or more inhabitants, a density of not less than 400 per sq. km, pronounced urban characteristics, and at least three-fourths of the adult male population employed in pursuits other than agriculture.

As per 2011 census, India has 468 towns with a population more than 0.1 million. Also, during the second half of the last century, the number of cities in India with a population of one million and above has steadily increased from 35 in 2001 to 52 in 2011, which is expected to further increase to 70 by 2025. Observing the historical trends of population growth in India (Table 1.2), the annual average growth rate in every decade has been positive since 1931. Similar positive trends can be seen for the growth in percentage of urban population to the total population since 1921 and which is expected to increase from 28% in 2001 to 58% by 2025. From the trends of urban population, similar trends can be observed for developing countries compared to developed countries (Figure 1.3).

Clearly, this growth of urban population in developing countries has a definite impact on travel demand and subsequently on urban mobility. It is clearly understood that cities are the economical contributors of the nation with its 50–60% contribution to GDP. However, the question is how can urban India survive with the estimated population as indicated above?

The most important factors common to India and other developing countries are population growth, increasing urbanization, rising motorization, and low per-capita income. The rapid growth of Indian cities has generated a correspondingly rapid

TABLE 1.2
Historical Growth of Population Growth

Year	Population (in Millions)	Density of Population per sq. km	Average Annual Exponential Growth Rate	Percentage of Urban Population to Total Population
1901	238.40	77	—	10.85
1991	252.09	82	0.56	10.29
1921	251.32	81	−0.03	11.18
1931	278.98	90	1.04	11.99
1941	318.66	103	1.33	13.86
1951	361.09	117	1.25	17.29
1961	439.23	142	1.96	17.97
1971	548.16	177	2.22	19.91
1981	683.33	216	2.20	23.33
1991	846.42	267	2.14	25.70
2001	1028.61	325	1.95	27.83
2011	1210.19	382	1.63	31.80
2021*	1369.13	475	1.24	34.71

Source: Census figures for different years. Figures for 2021 are estimated*

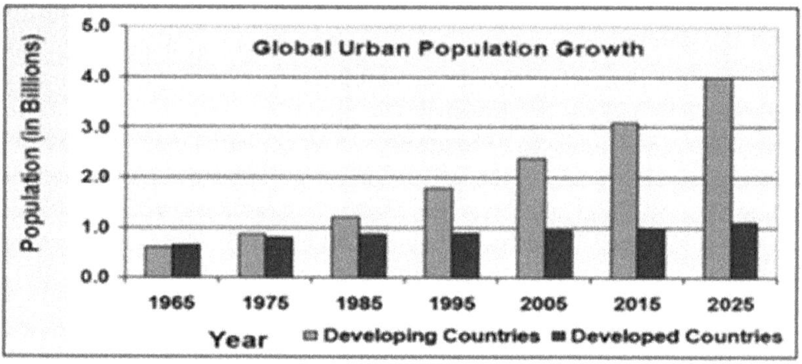

FIGURE 1.3 Population Growth in Developed and Developing Countries

Source: MoHUA (erstwhile MOUD)

growth in travel demand, overwhelming the limited transport infrastructure. The sharply increasing levels of MV ownership and use, in particular, have resulted in alarming levels of congestion, air pollution, noise, and traffic accidents. For most segments of the population, this has resulted in reduced mobility and accessibility over a period of time.

1.4 GROWTH OF URBAN SETTLEMENTS

In the case of India, census authorities adopted the following criteria for defining urban areas:

1. All places with a municipality, corporation, cantonment board, notified town area committee, or other such places
2. All places which satisfy the following criteria:
   a. A minimum population of 5000
   b. At least 75% of male working population engaged in nonagricultural pursuits.
   c. A density of population of at least 400 people per sq. km.
3. Besides the major project colonies, areas of industrial development, railway colonies, and important tourist destinations were also treated as urban though they might not fulfill the above criteria strictly.

Urban areas could be classified into different cities as per population. In India, many researchers, for purposes of infrastructure planning, have divided them into "6" categories. The number of urban centers and population in each category during 1991–2021 is presented in Table 1.3. In addition to the above classification, India also has more than 50 million-plus cities which pose traffic and transport problems and need special attention. The details of these cities are given in the next section.

TABLE 1.3
Distribution of Urban Settlements and Its Population

	1991		2001			2011			2021 (Estimated)	
Population Classes	No of Cities	Population (in 1000s)	No of cities	Population (in 1000s)	% Decadal Growth (1991–2001)	No of Cities	Population (in 1000s)	% Decadal Growth (2001–11)	No of cities	Population (in 1000s)
Less than 5000	198	663	192	668	0.7	424	1678	151.3	2794	18,800
5000–9999	739	5651	879	6658	17.8	1749	12,657	90.1	2774	40,420
10,000–19,999	1171	17,074	1346	19,458	14.0	1683	24,013	23.4	1942	57,810
20,000–49,999	944	28,688	1163	35,155	22.5	1373	41,833	19.0	1380	69,090
50,000–99,999	345	23,629	404	27,832	17.8	474	32,180	15.6	577	74,260
More than 100,000	300	140,067	372	196,348	40.2	468	264,900	34.9	1045	209,620
All classes	3697	215,772	4356	286,120	32.6	6171	377,261	31.9	9935	470,000

1.5 URBAN DEVELOPMENT IN INDIA

A meager 25% of India's population during 1991 lived in the 3700 areas of India that are classified as urban. Most of these urban areas are in places where there is prosperous agriculture which is in western, southern, and northwestern India though that is starting to change. Urban growth is surprisingly faster than India's rural growth even though there is a high amount of congestion in most cities. The reason for this exponential urban growth is due to the commercialization of the agricultural industry as well as the expansion of various other industries such as manufacturing and services. Some urban areas were comparable to major cities of the world in terms of urban population and density of population. The top five major urban centers of India and their population as per 2021 estimates are as follows:

1. Mumbai: 23.9 million
2. Delhi: 21.2 million
3. Kolkata: 18.2 million
4. Chennai: 11.2 million
5. Bengaluru: 11.1 million

All these cities are experiencing traffic congestion during most of the daytime in one locality or another.

1.6 MIGRATION TO URBAN AREAS

Rural households' main occupation is agriculture with 75–80% deriving a large part of the income through agriculture-related activities. About 10–15% of rural households generate secondary income through dairy (milk)-related activities. About 10–15% of households only generate income through other means such as employment, manufacturing, and trade.

Urban centers provide better opportunities for education, health, employment, and social interaction. Due to these better opportunities, migration to urban areas is unstoppable and this fact needs to be considered while planning transport infrastructure in an appropriate manner. This migration has led to slums in various cities. This slum population is also increasing in almost all urban centers. Table 1.4 provides migration details to urban areas. A common perception is that explosive rural-to-urban migration is the primary cause for the state of India's cities. This is not borne out by the evidence. For the last 30 years, migration has contributed about a fifth of the urban population, natural urban population growth contributed about 60%, and the rest about equally split between new town formation (because of reclassification) and urban boundary expansion or sprawl. One important facet of study on population is the study of migration arising out of various social, economic, or political reasons. For a large country like India, the study of movement of population in different parts of the country helps in understanding the dynamics of the society better. At this junction in economic development, in the country, especially when many states are undergoing faster economic development, particularly in areas such as manufacturing, information technology, or service sectors, data on migration profile of population has become more important.

TABLE 1.4
Migration Details from Rural to Urban

Year	Migration from Rural to Urban Areas (in Million)	Decadal Percent Growth
1951	NA	-
1961	66	-
1971	68.2	3.33
1981	81	18.77
1991	80.9	–0.12
2001	98.3	21.51
2011	59	–39.98

Source: MoHUA

TABLE 1.5
Million-Plus Cities in India

Rank	City	Population (in Million)				Percent Increase in the Decade	Percent Increase in the Decade	Percent Increase in the Decade
		1991	2001	2011	2021*	(1991–2001)	(2001–11)	(2011–21)
1	Mumbai	12.57	16.37	18.39	23.90	30.23	12.34	29.96
2	Kolkata	10.92	13.22	14.06	18.28	21.06	6.35	30.01
3	Delhi	8.38	12.79	16.35	21.22	52.63	27.83	29.78
4	Chennai	5.36	6.42	8.65	11.20	19.78	34.74	29.48
5	Hyderabad	4.28	5.53	7.68	10.27	29.21	38.88	33.72
6	Bengaluru	4.09	5.69	8.52	11.25	39.12	49.74	32.04
7	Ahmadabad	3.3	4.52	6.36	8.21	36.97	40.71	29.08
8	Pune	2.49	3.75	5.06	6.99	50.6	34.93	38.14
9	Kanpur	2.11	2.69	2.92	3.15	27.49	8.55	7.87
10	Nagpur	1.66	2.12	2.5	2.94	27.71	17.92	17.60

Source: Government of India Ministry of Statistics and Programme Implementation Central Statistics Office Social Statistics Division. Census 2001

For the purpose of providing an efficient public transport infrastructure, the urban centers are classified into category I to category V depending upon population. In total, there are more than 50 million-plus cities in terms of population in India. The growth pattern of top 10 million-plus cities from 1991 to 2011 is presented in Table 1.5.

1.7 GROSS DOMESTIC PRODUCT

GDP is the market value of all officially recognized final goods and services produced within a country in a given period. GDP per capita is often considered an indicator of a country's standard of living.

Urban areas contribute to the national income at a much higher rate compared to rural population. Contribution to national wealth by urban as well as rural population in the past six decades is presented in Table 1.6 and Figure 1.4. There was a rapid increase in the GDP of India from 2001 to 2011. This increase is more than three times during this 10-year period. Similarly, it is observed that during the 10-year period (2011–21), the increase is more than 1250 billion.

It may be noted that the contribution of urban population per million is increasing at a faster rate compared to the contribution made by rural population. Currently, a million urban population contributes 2.5 times that of rural population. This underlines the need and rationale for providing adequate transport infrastructure to enhance its contribution to GDP of the nation. Figure 1.5 shows the GDP values of India from 1950 to 2021.

1.8 URBAN TRANSPORTATION STATUS IN DEVELOPING COUNTRIES

Many countries, including India, are experiencing rapid urbanization. This uncontrolled development has many important implications in terms of social, environmental, and economic repercussions at a country level. In terms of transport, there are two major policy decisions that must be acknowledged: the first is the decision as to the priorities of the automobile industry. The second is the decision to improve the urban public transport network. These two decisions, apparently contradictory, must be developed coherently to achieve an optimum mix of improved public and private transport situations for its cities, while not hampering its industrial development. According to a World Bank study, the urban transport problems have been associated with globalization, urbanization, fiscal decentralization, and an economic transition. There are also growing concerns such as suburbanization of homes and consequent longer trips. In addition, it should be noted that there is a lack of technical capacity and practical experience in strategic planning in transport. Finally, there is little

TABLE 1.6
Percentage Contribution of GDP by Rural and Urban Areas

Year	1951	1981	1991	2001	2011	2021
GDP of India (in billion USD)	NA	196.1	274.8	492.4	1873	3150
Percentage Contribution of Urban Area	30	48	55	60	70	63
Percentage Contribution of Rural Area	70	52	45	40	30	37

Source: GDP of India (1951–2021)

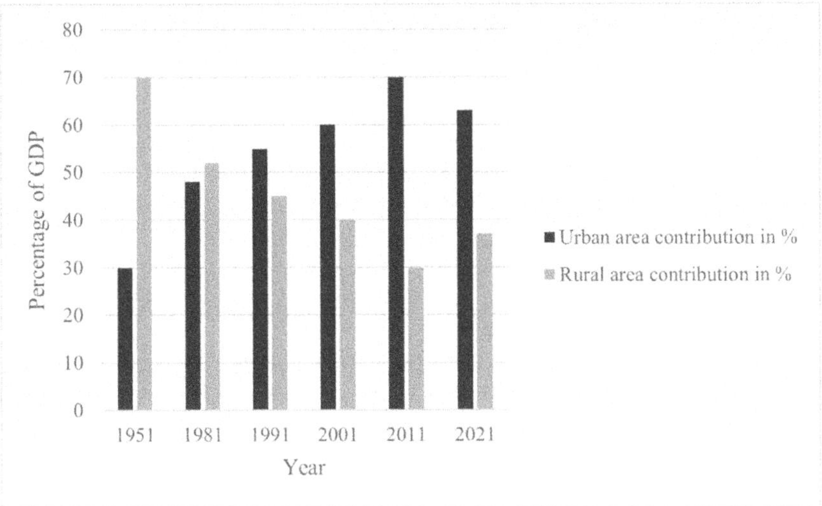

FIGURE 1.4 Percentage GDP Contributions by Urban and Rural Areas

Source: GDP of India (1951–2021)

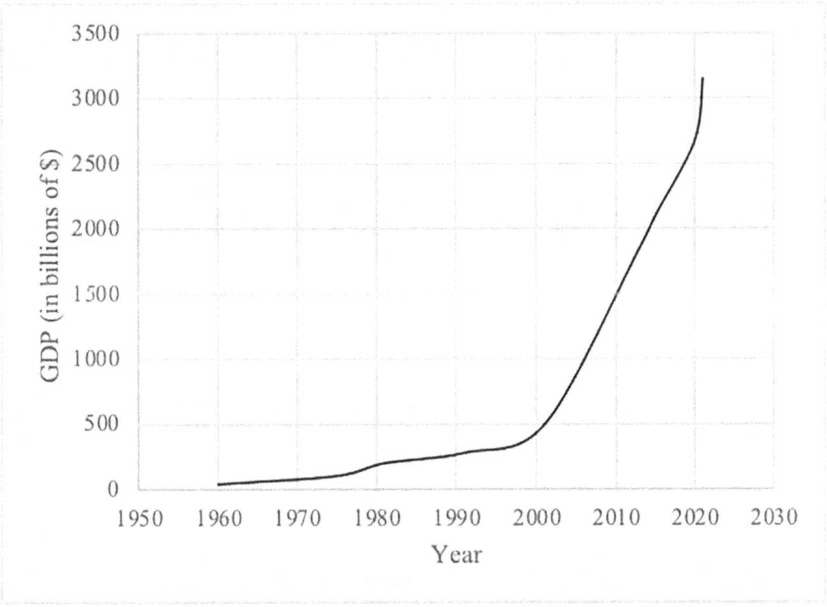

FIGURE 1.5 GDP Values of India From 1951 to 2010

Source: GDP of India (1950–2010)

public participation in planning processes, and there is a tendency toward technological improvements rather than modified policies, priorities of sustainable means of transport, or demand management techniques to improve transportation. The analysis of data and information available on developing countries clearly suggests that the current systems and trends of urban transportation, with respect to both mobility and safety, are not sustainable. The cities of developing countries are typically characterized by high-density urban areas, absence of proper control on land use, lack of proper roads and parking facilities, poor public transport, lack of road user discipline, etc. This results in transportation problems, namely accidents, congestion, and pollution, taking a very different and much severe shape in developing countries than those in cities of developed countries.

The urban transport characteristics of India are proposed to be captured by studying the transport dynamics of cities with million-plus population. As per 2001 census, these cities together accounted for about 33% of the Indian urban population and about 28.5% of the total vehicles in 2005. The number of million-plus cities in India increased to 52 as per census of 2011 and is estimated to have increased to 65 by 2021. India's urban population is expected to increase from 286 million in 2001 to 534 million in 2026 (38%). The urban transport conditions are shown in Figure 1.6.

There is an urgent need to improve urban infrastructure to achieve the objectives of economic development. However, most of the cities in India have inadequate transport infrastructure. Urban transport is one of the major problem areas, affecting the mobility of people and accelerated economic growth of the urban centers of India. These problems are due to the prevailing imbalance in modal split, inadequate transport infrastructure and its suboptimal use, and an absence of proper integration between land use and transport planning.

FIGURE 1.6 Glimpse of Transport Condition in India

In addition to million-plus cities, India has more than 80 urban areas that are already grappling with many of the challenges mentioned above. These challenges arise because government and city planners have approached urban planning more as a knee-jerk reaction than from a strategic planning point of view to guide future growth. They have failed to understand the need to integrate urban transport development with city growth and, as a result, today we are faced with a huge challenge in providing a sustainable transport system to a population that is engaged in a variety of economic activities. Though an array of transport modes are available, Indian cities lack the necessary infrastructure to cater to an increased travel demand.

An associated problem has been the declining reliance on public transport. As a result, the number of personal vehicles in Indian cities is growing at an unforeseen rate. Though significant improvements in public transport have been made in certain specified locations of some cities, these trends are still visible in many of the metropolitan cities including Delhi and its satellite towns like Noida, Gurgaon, and Ghaziabad. For instance, the number of personal vehicles per 1000 population in Delhi has expanded by about 4 times in the last three decades, whereas the number of buses per 1000 population has increased only 2.3 times.

According to available data for all India, the share of public transport buses declined to 0.9% of total vehicle population in 2008 from 11.1% in 1951. Several urban transport studies also reveal that cities without good mass transit systems showed a higher growth rate in vehicular population as compared to those with mass transit systems. Average vehicular speeds in many of these cities were as low as 10 km per hour during peak hours. The root cause of this problem is the failure to adapt the urban public transport system to the unique needs of these emerging cities. As a result, many cities in India today are faced with urban transport challenges like the following:

1. Mixed traffic flow and mixed land use
2. Severe congestion, pressure on existing public transport networks, and increased travel demand
3. Excessive delays, low speeds, and high travel times
4. Increased fuel consumption and air pollution
5. Safety issues
6. Exacerbation of problems due to rapid urbanization
7. Poor public transport and infrastructure deficiencies
8. Exponential growth in vehicle ownership

These problems are likely to intensify, if not multiply. With the transport sector already contributing to over 9% of India's total GHG emissions, the increased use of private means of transportation will adversely impact energy use and environmental quality.

SUMMARY

This chapter examines India's demographic shifts, urbanization, and transportation challenges. It begins with an overview of the country's geography, population, and economic growth, emphasizing the role of transportation in urban development. It then explores demographic changes since independence, highlighting population growth and rural-to-urban migration. The growth of urban settlements and urban development are discussed in relation to industrialization, infrastructure, and government policies. The chapter also examines migration trends, their impact on city infrastructure, and the contribution of urban areas to GDP. Finally, it addresses urban transportation in developing countries, focusing on rising motorization, congestion, and emissions. The need for sustainable transport solutions is emphasized to ensure efficient and eco-friendly urban mobility.

2 Different Public Transport Modes, Their Characteristics, and Operating Elements

2.1 MODES OF TRANSPORT

Modes of transport or transport modality is a term used to describe the substantially different means of transportation. The ability of people and goods to move from one place to another is fulfilled using one or another mode of transport.

The most dominant modes of transport are air transport, land transport, and water transport. Land transport includes road, rail, and off-road transport. Off-road transport consists of pipelines, cable transport, etc. Animal and human-powered transport are sometimes regarded as modes, but these normally also fall into the abovementioned categories. Each mode of transport has a fundamentally different technological solution, and some require a separate environment. Each mode has its own infrastructure, vehicles, and operations, and often has its unique regulations. Each mode also has separate subsystems like propulsion, suspension, control, guidance, structure, and support. Transport that uses more than one mode is described as intermodal. Transportation that carries around many people and can be used by the public is known as mass transportation.

2.2 TRANSPORT SYSTEMS DEFINITIONS AND CLASSIFICATIONS

Urban transportation modes and operational concepts can be classified in different ways. Some of the classifications are interdependent. For example, modes are often identified with system technology only, but the same is strongly influenced by the characteristics of rights-of-way (ROWs) and operations. The following subsections discuss these classifications.

2.2.1 Classification by Type of Usage

There are three basic categories of transportation by type of operation and usage:

1. Personal or private transportation
2. For hire or intermediate public transportation (IPT)
3. Public or mass transportation

The main characteristics, typical modes, and operating domains of these categories are shown in Tables 2.1 and 2.2.

DOI: 10.1201/ 9781003740247-2

TABLE 2.1
Classification of Urban Passenger Transportation by Type of Usage

Characteristics/Usage Type	Personal	Intermediate	Public
Common designation	Private transportation	Paratransit	Transit
Service availability	Owner	Individuals, group	Public
Service supplier	User	Carrier	Carrier
Route determination	User (flexible)	User (carrier)	Carrier (fixed)
Time-schedule determination	User (flexible)	User (carrier)	Carrier (fixed)
Cost price	User absorbs	Fixed rate	Fixed fare

Source: V. R. Vuchic, 2005—Urban Transit

TABLE 2.2
Classification of Urban Passenger Transportation by Carrier Type

	Individuals				
Carrier Type	Personal		Intermediate		Group
Modes	Walking, bicycle motorcycle automobile	Carpools, van-pools	Rental car. cars, sharing taxi auto-rickshaws, cycle rickshaws, hand-pulled rickshaw, horse-pulled carriages	Dial-a-ride, charter bus	Street-transit (bus, trolleys, streetcar), semi-rapid transit (bus-rapid transit, light rail transit (LRT)), rapid transit (rail and metros), specialized modes
Optimum (but not exclusive) domain of operation area density	Low–medium	Origin: low; destination: high	Any		High–medium
Routing	Dispersed	Radial	Dispersed		Concentrated (radial), ubiquitous
Time	Off-peak	Peak only	All times		Peak, daily hours
Trip purpose	Recreation, shopping, business, other	Work only	Business special services		Work, school, business, social, other emergency

Source: V. R. Vuchic, 2005—Urban Transit

2.2.2 Personal Transport Modes and Their Characteristics

Personal transportation consists of privately owned vehicles operated by owners for their own use, usually on publicly provided and operated streets. Private automobiles are the most common mode in the category of personal transport, but motorcycles, bicycles, and, of course, walking also belong to this category. The following subsections discuss different personal transport modes and their characteristics.

2.2.2.1 Walking or Jogging

Walking is the most energy-efficient, healthy, and sustainable mode of transport. It is also the most universal mode. Everyone is a pedestrian for part of their trip even if they are driving or using a bus or rail transit. Hence, walking needs to be the primary focus of any plan or policy related to urban transport in Indian cities. It has been observed that people do not prefer walking or cycling even for short distances. This is predominantly because of the lack of pedestrian infrastructure and walkable street networks.

Currently, the share of public transport trips in major urban areas in India is between 25% and 30%. As every public transport user is also a pedestrian at the time of access and egress, walking trips may constitute 50–60% of total trips. Despite the high share of walk trips, the transport infrastructure does not include any facilities for this mode. With the exception of a few cities, street networks in most cities do not provide even minimum width (2 m) foot paths.

If our cities are to sustain walking as a viable mode of transport (even for short/medium trips between 1 and 3 km), provision of pedestrian infrastructure needs to be an immediate priority. Safety and comfort of pedestrians are equally important. People will not walk in unsafe environments, which are often the after effects of automobile-oriented street designs. Hence, designing streets where walking is prioritized is important along with building regulations that also prioritize access and egress by pedestrians. As per Rastogi (2007), the per hour per meter capacity of public transport modes like busways, suburban rail, and surface rapid rail ranges between 4000 and 8000. It is 3500 for walking and 1500 for cycling. The similar values for other motorized vehicles usually range between 100 and 400 meter per hour.

2.2.2.2 Bicycles and Motorcycles

After walking, cycling (Figure 2.1) is the second-most efficient and effective mode of transportation for short to moderate distances. There are many health as well as environment-related benefits of walking and cycling. The bicycle, which is also known as green transport, is often an important consideration to improve the urban transport problems. Many countries recognize the potential of cycling to grow as a mode of transport and act as an alternative to car use to create more sustainable urban futures.

Currently in India, cycling is considered to be the mode of transportation that is used by lower income groups of society who cannot afford any other means of transport. Higher income groups of people prefer using cars or two-wheelers even for short distances. This reality needs to change and use of these modes needs to be brought into the day-to-day lifestyle of all sections of society if our future is to be sustainable.

FIGURE 2.1 Glimpse of a Cycle

Currently, most of the street space allocation and provision of infrastructure cater to private vehicle users. To make cycling an urban transport priority, one of the most important tasks is to create a high-quality cycling infrastructure. This means providing dedicated and segregated cycle lanes on all major arterials along with proper cycle crossing, cycle parking, and other facilities at regular intervals.

Several cities in the world have been successful at promoting the use of cycling by having a strong public bicycle sharing program. Such programs have been implemented in many cities including New York; Washington, DC; Guangzhou; Hangzhou; London; and Paris. Some cities in India (Bengaluru, Delhi, etc.) have initiated pilot projects along similar lines. Public bicycle sharing not only helps promote cycling but also helps sensitize users of other modes to the needs of cyclists. Moreover, integrating the provision of cycling infrastructure with the existing and planned public transport modes is important to link the trip chains. It also allows a larger group of people to access public transport at a very low/no cost and use it for longer trips.

As per Rastogi (2007), the comparative overview of space requirements by different modes with respect to the bicycle requirements indicates that only light rail transit (LRT) and walk consume less bicycle space equivalent, whereas in the case of cars it is quite high. In terms of per person, the road space required by a car is 120 sq. m.; for a bus, it is 12 sq. m.; by rail, it is 7 sq. m.; for a bicycle, it is 9 sq. m,; and by pedestrian, it is 2 sq. m.

2.2.2.3 Automobiles

An automobile, auto car, motorcar, or car is a wheeled motor (Figure 2.2) vehicle used for transporting passengers, which also carries its own engine or motor. Most definitions of the term specify that automobiles are designed to run primarily on roads, to have seating for one to eight people, to typically have four wheels, and to be manufactured principally for the transport of people rather than goods.

FIGURE 2.2 Glimpse of an Automobile

The year 1886 is regarded as the year of birth of the modern automobile—with the Benz Patent-Motorwagen, by German inventor Carl Benz. Motorized wagons soon replaced animal-drawn carriages, especially after automobiles became affordable for many people when the Ford Model T was introduced in 1908.

The term "motorcar" has formerly also been used in the context of electrified rail systems to denote a car which not only functions as a small locomotive but also provides space for passengers and baggage. These locomotive cars were often used on suburban routes by both interurban and intercity railroad systems.

It was estimated in 2010 that the number of automobiles had risen to over 1 billion vehicles, up from 500 million in 1986. In 2024, there were estimated 1.45 billion cars in the world. The numbers are increasing rapidly, especially in China and India.

The automobile sector is divided into four segments: two-wheelers (mopeds, scooters, motorcycles, electric two-wheelers), passenger vehicles (passenger cars, utility vehicles, multipurpose vehicles), commercial vehicles (light and medium-heavy vehicles), and three-wheelers (passenger carriers and good carriers).

The automobile industry is one of the key drivers of economic growth of the nation. Since the delicensing of the sector in 1991 and the subsequent opening up of 100% FDI through automatic route, the Indian automobile sector has come a long way. Today, almost every global auto major has set up facilities in the country.

The world standings for the Indian automobile sector, as per the Confederation of Indian Industry, are as follows:

1. Largest three-wheeler market
2. Second largest two-wheeler market
3. Tenth largest passenger car market
4. Fourth largest tractor market
5. Fifth largest commercial vehicle market
6. Fifth largest bus and truck segment

2.2.3 Intermediate Public Transport

Commonly designated as paratransit, it is a transportation service provided by an operator and available to all parties who meet the conditions of a contract for carriages (i.e., pay prescribed fares or rates), but which is adjustable in various degrees to the individual user's desires. Most paratransit modes do not have fixed routes and schedules. Taxi, dial-a-ride, and auto-rickshaws are major modes.

Paratransit is an alternative mode of flexible passenger transportation that does not follow fixed routes or schedules. Typically, minibuses are used to provide paratransit service, but share taxis and jitneys (in China) are also important providers.

Paratransit mode routes and schedules that change with requests to individual users are called as demand-responsive. When the difference is pointed out and transit is pointed, that transit is described as fixed-route or fixed-schedule service.

Paratransit services may vary considerably depending on the degree of flexibility they provide their customers. At their simplest, they may consist of a taxi or small bus that will run along a defined route and then stop to pick up or discharge passengers on request. At the other end of the spectrum of fully demand-responsive transport, the most flexible paratransit systems offer on-demand call-up, door-to-door service from any origin to any destination in a service area. Paratransit services are operated by public transit agencies, community groups, not-for-profit organizations, and for-profit private companies or operators.

2.2.3.1 Taxicab

A taxicab, also known as a taxi or a cab, is a type of vehicle for hire with a driver, used by a single passenger or a small group of passengers often for a non-shared ride. A taxicab conveys passengers between locations of their choice (Figure 2.3).

In modes of public transport, the pick-up and drop-off locations are determined by the service provider, not by the passenger, although demand-responsive transport and share taxis provide a hybrid bus/taxi mode. There are four distinct forms of taxicab, which can be identified by slightly differing terms in different countries:

1. Hackney carriages, also known as public hire, hailed, or street taxis, licensed for hailing throughout communities

FIGURE 2.3 Glimpse of a Taxi Cab

2. Private hire vehicles, also known as minicabs or private hire taxis, licensed for prebooking only
3. Taxi-buses, also known as jitneys, operating on the present routes typified by multiple stops and multiple independent passengers
4. Limousines, specialized vehicle licensed for operation by prebooking.

Although types of vehicles and methods of regulation, hiring, dispatching, and negotiating payment differ significantly from country to country, many common characteristics exist.

2.2.3.2 Auto-Rickshaws

Autorickshaws are a common means of intermediate transportation in many countries in the world (Figure 2.4). Also known as a three-wheeler, samosa, tempo, tuk-tuk, trishaw, auto–rickshaw, autorik, bajaj, rick, tricycle, mototaxi, babytaxi, or lapa in popular parlance, an auto–rickshaw is a usually three-wheeled cabin cycle for private use and as a vehicle for hire.

It is a motorized version of the traditional pulled rickshaw or cycle rickshaw. Auto–rickshaws are an essential form of urban transport in many developing countries, and a form of novelty transport in many Eastern countries.

2.2.3.3 Airport Taxi

Airport taxi is one of the IPT modes in major cities which provide good services for the people to travel from their origin to an airport or from an airport to a destination (Figure 2.5). There are many private cab services in major cities. For example, in Bengaluru, there are Meru cabs, Easy cabs, and Hire cabs which play an important role as the paratransit modes.

2.2.3.4 Hand-Pulled Rickshaws

Around 1880, rickshaws appeared in India, first in Shimla. At the turn of the century, it was introduced in Kolkata (Calcutta), India, and in 1914, it was a conveyance

FIGURE 2.4 Glimpse of an Auto-Rickshaw

FIGURE 2.5 Glimpse of Taxis

FIGURE 2.6 Hand Pulled Rickshaws

for hire. A pulled rickshaw (or rickshaw) is a mode of human-powered transport by which a runner draws a two-wheeled cart which seats one or two people (Figure 2.6). In recent times, the use of human-powered rickshaws has been discouraged or outlawed in many countries due to concern for the welfare of rickshaw workers. Pulled rickshaws have been replaced mainly by cycle rickshaws and auto–rickshaws.

2.2.3.5 Horse-Pulled Carriage

A carriage is a wheeled vehicle for people, usually horse-drawn (Figure 2.7). It excludes litters (palanquins) and sedan chairs because they are wheelless vehicles. The carriage is especially designed for private passenger use and for comfort or elegance, though some are also used to transport goods. It may be light, smart, fast or heavy, large, and comfortable. Carriages normally have suspension using leaf springs, elliptical springs (in the nineteenth century), or leather strapping. A public passenger vehicle would not usually be called a carriage. Terms for such public passenger vehicles include stagecoach, chara banc, and omnibus. Working vehicles such as the (four-wheeled) wagon and (two-wheeled) cart share important parts of the history of the carriage, such as a fast (two-wheeled) chariot.

FIGURE 2.7 Horse Pulled Rickshaws

FIGURE 2.8 Glimpse of an Ox Wagon

2.2.3.6 Ox Wagon

An ox wagon or bullock wagon is a four-wheeled vehicle pulled by oxen (draught cattle) (Figure 2.8). It was a traditional form of transport, especially in Southern Africa but also in New Zealand and Australia. Ox wagons were also used in the United States. The first recorded use of an ox wagon was around 1670, but it continues to be used in some areas up to modern times.

FIGURE 2.9 Cycle Rickshaws

2.2.3.7 Cycle Rickshaw

Cycle rickshaw is a small-scale local means of transport; it is also known by a variety of other names such as bike taxi, velotaxi, pedicab, bikecab, cyclo, beca, becak, trisikad, or trishaw (Figure 2.9). As opposed to rickshaws pulled by a person on foot, cycle rickshaws are human-powered by pedaling. They are a type of tricycle designed to carry passengers on a hire basis. Cycle rickshaws are widely used in major cities around the world, but most commonly in cities of South, Southeast, and East Asia.

2.3 TRANSIT MODES' CHARACTERISTICS

Transit mode is mainly considered on three basic characteristics

1. ROW category
2. System technology
3. Type of service

Transit modes vary with each one of these characteristics. Contrary to the common belief that technology mostly determines modal characteristics, the ROW category has a major influence on both performance and costs of modes.

2.3.1 ROW Category

A transit way or ROW is the travel way or strip of land on which the transit vehicles operate. There are three basic ROW categories distinguished by the degree of their separation from other traffic.

1. Category C represents surface streets with mixed traffic. Transit may have preferential treatment, such as reserved lanes separated by lines or special signals or travel mixed with other traffic.
2. Category B includes ROW types that are longitudinally physically separated by curbs, barriers, grade separation, and the like from the other traffic but with grade crossings for vehicles and pedestrians, including regular street intersections. This category is most frequently used for the LRT system.

3. Category A is a fully controlled ROW without grade crossings or any legal access by other vehicles or people. It is also referred to as grade-separated, private or exclusive ROW, and it can be a tunnel, an aerial structure, or at grade level.

2.3.2 System Technologies

Technology of transit modes refers to the mechanical features of their vehicles and ways. The four most important features are defined here:

1. Support is a vertical contact between vehicle and riding surface, which transfers the vehicle weight and traction force. The most common types are rubber tires on concrete, asphalt, or other surface and steel wheel on steel rail.
2. Guidance refers to the means of lateral vehicle guidance. Highway vehicles are steered (by the driver) and their lateral stability is provided by wheel/support adhesion. Rail vehicles are guided by flanges and the conical form of the wheel surfaces.
3. Propulsion refers to the type of propulsion unit and method of traction or transferring acceleration/deceleration forces. Its major components are type of propulsion unit, method of transferring tractive forces.
4. Control is the means of regulating the travel of one or all vehicles in a system. The most important control is for longitudinal spacing of the vehicles, which may be manual-vision, manual-signal, fully automatic, or various combinations of these.

TABLE 2.3
Classification of Urban Public Transportation Modes by ROW Category and Technology

ROW/ Technology	Highway Driver Steered	Rubber-Tired Guided, Partially Guided	Rail	Specialized
C	Paratransit, shuttle bus, regular bus (RB, on street)	Trolley bus	Streetcar/tramway/ cable car	Ferryboat hydrofoil
B	Bus rapid transit (BRT)	Guided bus	LRT	(cog railway)
A	Bus on bus bay only	Rubber-tired metro, rubber-tired monorail automated guided transit (AGT) PRT	LRT, rapid rail, rapid transit/metro regional/commuter rail monorail, schwebebahn	Cog railway, funicular, aerial tramway

Source: V. R. Vuchic, 2005—Urban Transit

2.3.3 Types of Service

There are many different types of services. They can be classified into groups by three characteristics:

1. By the types of routes and trips served
2. By stopping schedule or type of operation
3. By time of operation

2.4 PUBLIC TRANSPORT MODES AND CLASSIFICATION OF MASS TRANSIT MODES

Mass transportation or transit is a common type of urban transport. Such transport systems have fixed routes and schedules, available for use by all people who pay the established fare. The most common representatives are bus, LRT, and rapid transit or metro, but there are several other modes.

Public transport (North American English: public transportation or public transit) is a shared passenger transport service which is available for use by the public, as distinct from modes such as taxicab, carpooling, or hired buses which are not shared by strangers without private arrangement.

Public transport modes include buses, trolleybuses, trams, trains, and rapid transit (metro/subways/undergrounds/etc.). Public transport between cities is dominated by airlines, coaches, and intercity rail. High-speed rail networks are being developed in many parts of the world.

Most public transport runs to a scheduled timetable with the most frequent services running to a headway. Share taxi offers on-demand services in many parts of the world and some services will wait until the vehicle is full before it starts. Paratransit is sometimes used in areas of low demand and for people who need door-to-door service.

Urban public transport may be provided by one or more private transport operators or by a transit authority. Public transport services are usually funded by government subsidies; fares are charged to each passenger. Services are normally regulated and possibly subsidized from local or national tax revenue. Fully subsidized, zero-fare (free) services operate in some towns and cities.

For historical and economic reasons, there are differences internationally regarding the use and extent of public transport. While countries in the world tend to have extensive and frequent systems serving their old and dense cities, many cities of the new world have more sprawl and much less comprehensive public transport.

Common-carrier urban passenger transport mode is known as transit, mass transit, or mass transportation. These are transport systems with fixed routes and schedules available for use by all people who pay the established fares. Transit is described as a fixed route, fixed schedule service. The best-known classification of transit modes is into three generic classes based mostly, but not entirely, on right-of-way (ROW) type [Gray and Hoel (1979)]. They are street transit or surface transit, semi-rapid transit, and rapid transit or mass rapid transit systems.

The capacity of these public transport modes can be calculated as follows:

For bus:

$$\text{Passenger per hour per direction} = \frac{\text{Buses}}{\text{hour}}\left[\frac{\text{seats}}{\text{buses}} * \frac{\text{passengers}}{\text{seat}} + \frac{\text{standing area}\left(\text{m}^2\right)}{\text{bus}} * \frac{\text{standees}}{\text{m}^2}\right] \quad (2.1)$$

For rail systems:

$$\text{Passenger per hour per direction} = \frac{\text{Trains}}{\text{hour}} * \frac{\text{Cars}}{\text{Trains}}\left[\frac{\text{seats}}{\text{car}} * \frac{\text{passengers}}{\text{seat}} + \frac{\text{standing area}\left(\text{m}^2\right)}{\text{bus}} * \frac{\text{standees}}{\text{m}^2}\right] \quad (2.2)$$

2.4.1 Street Transit or Surface Transit

Street transit modes are operated on streets with mixed traffic (i.e., ROW category C); its reliability is often low because of various interferences and its speed is lower than the speed of traffic flow, owing to the time lost at passenger stops. This class includes the following modes:

1. *Minibus:* These are smaller diesel- or petrol-driven buses and have their present use largely in private operation. They are useful for narrow, crowded streets, where large buses have a problem in maneuvering, and also for low-demand routes where the provision of large buses may prove to be uneconomical (Figure 2.10).
2. *Regular Bus (RB):* It consists of single-decker buses operating along fixed routes on fixed schedules. Buses comprise by far the most widely used

FIGURE 2.10 Minibus

transit mode. The more travel demand is concentrated along corridors, the more advantageous the RB becomes (Figure 2.11). The most typical bus services are street transit routes, which may represent the entire transit network (small and most medium-size cities) or supplementary and feeder services to rail networks.

3. *Double-Decker Bus:* This is a higher capacity bus, with a design capacity of about 114 and crush capacity of 130. The bus is of size similar to an RB, though of course higher, with a more powerful engine. It is however slower for two reasons, namely power-to-weight ratio and longer stop times. It is not as economical (mainly due to its high capital cost) as an RB (Figure 2.12). In addition, it has a disadvantage that all flyovers, elevated ROW, bridges, etc. need clearance of 5.5 m instead of 4–4.5 m.
4. *Articulated Bus:* These buses are usually 55 ft. or more in length with two connected passenger compartments that bend at the connecting point when

FIGURE 2.11 Regular Bus

FIGURE 2.12 Double-Decker Bus

FIGURE 2.13 Articulated Bus

the bus turns a corner (Figure 2.13). It is not recommended, though it has a higher capacity, because its extra length creates its own problems under Indian traffic conditions. The driver cannot be as aware of the rear of the bus and cornering is more difficult. As important is the fact that at stops the driver is not able to pull in correctly even with the standard and takes up two lanes. In addition, the bus stop lengths would need to be altered.

5. *Express Bus:* Express bus service is provided by fast, comfortable buses on long routes with widely spaced stops. Its reliability of service depends on traffic conditions along the route (Figure 2.14). These services are mostly run on intercity on interstate operations.
6. *Trolley Bus (TB):* Trolley buses are the same vehicles as buses except that they are propelled by an electric motor and obtain power from two overhead wires along their route (Figure 2.15). The advantages the TB offers include higher riding quality (smooth vehicle motion) and excellent environmental features (extremely low noise, no exhaust). It is always more expensive than the standard bus because at low demand levels its infrastructure cost makes it more expensive, and at high demand levels, it cannot meet the demand, being limited to two lanes. However, as these factors are not reflected in the operators' revenues, financial problems of transit agencies have often led to substitution of buses for TBs.
7. *Streetcar (SCR) or Tramway:* SCRs are electrically powered rail transit vehicles operating mostly on streets as shown in Figure 2.16. Their tracks and

FIGURE 2.14 Express Bus

FIGURE 2.15 Trolley Bus

distinct vehicles give transit service a strong identification. When compared with buses, the SCRs have more comfortable rides, quieter and pollution-free operation, better vehicle performance, higher labor productivity (large vehicles), higher line capacity, but higher investment cost, less reliable street operation unless transit enjoys priority treatment, less flexible operation, higher maintenance, and greater impedance of other traffic. A tram is by far the most expensive street transit mode, as well as with very low capacity.

8. *Guided Bus:* Guided buses free the driver from the task of steering the bus while operating in a section of route equipped with guided-bus infrastructure (Figure 2.17). The guidance could be lateral, central, or electronic. The theoretical maximum capacity quoted for guided buses is 12,000 pphpd

FIGURE 2.16 Tramway

FIGURE 2.17 Guided Bus

(passengers per hour per direction), but the realized capacity of the guided busway system as observed in Adelaide, Australia, is around half the theoretical capacity just quoted.

9. *Battery-Operated Bus:* Battery-operated buses (Figure 2.18) have low range and speed capability and cannot be used on a large scale in a city. However, they offer the following advantages as a low capacity (500–1000 pphpd) people mover in a congested area:
   a. Economic at low speeds with frequent start/stops
   b. Low noise
   c. Small in size, so easy to maneuver.

FIGURE 2.18 Battery Operated Bus

TABLE 2.4A
System Parameters for Street Transit

	Minibus	Standard Bus	Double-Decker Bus	Trolley Bus	LRV (Tram)
L*W (m)	6.6*2.3	9.7*2.5	9.1*2.4	11*2.5	11*2.5
Turning radius (m)	7	11	12	22	15
Design (seats)	30 (20)	76 (35)	114 (70)	75 (40)	75 (40)
Crush	40	100	130	110	110
PCU	2	3	3.5	4[1]	5[1]
Acceleration (m per second^2)	0.8	0.4	0.4	1.2	0.6
Cruising speed (kmph)	60	60	50	60	50
Average speed (kmph)[2]	20	17	15	20	15
Life (years)	8	[illegible]	[illegible]	10	[illegible]
Cost (Rs. million) 1986 level	0.2	0.3	0.6	0.7	0.7
Energy source	D/P	D	D	E	E
Pollution (kg/000 veh.km at avg. speed)	35	38.05	38.05	0	0

[1] For reason of length as well as lack of maneuverability;
[2] at average congestion

Source: Compiled from GOI (1987)

They should be considered seriously for short routes and low-demand routes in city areas.

Tables 2.4a and 2.4b give the system parameters for street transit and capacity of transit system.

TABLE 2.4B
Capacity of Street Transit System

No. of Lanes per Direction	1(Single)	1	1	2(double)	2
Bus type	Large urban single decker	High-capacity double decker	Articulated	Large urban single decker	High-capacity double decker
Overtaking possible	No	No	No	Yes	Yes
Buses/hour/direction (normal breaking)	85 (headway 42 seconds)	85 (headway 42 seconds)	80 (headway 40 seconds)	240 (headway 15 seconds)	240 (headway 15 seconds)
Pphpd (normal load)	6460	9690	9440	18,240	27,360
Pphpd (crush load)	8500	11,050	12,000	24,000	31,200

Compiled from GOI (1987)

2.4.2 Semi-Rapid Transit

Semi-rapid transit consists of modes utilizing mostly ROW category B [i.e., ROW types which are longitudinally physically separated (by curbs, barriers, grade separation, etc.) from other traffic, but with grade crossings for vehicles and pedestrians, including regular street intersections]. This class includes the following modes:

1. *Semi-Rapid Buses (SRBs):* SRBs are regular or high-performance buses operating on routes that include substantial sections of ROW category B. Performance of such systems depends greatly on proportion and locations of separated ROW sections, ROW types, and types of operation.
2. *LRT:* LRT is a mode utilizing predominantly reserved, but not necessarily grade-separated ROW. Its electrically propelled rail vehicles operate singly or in trains.

LRT provides a wide range of LOS and performance characteristics. When compared with SRB on the corresponding alignments, LRT is characterized by easier securing of B or A ROW, stronger image and identity of lines (rail technology), more spacious vehicles, higher passenger attraction, low noise, no exhaust, better vehicle performance due to electric traction, higher system performance, ability to operate in tunnels, ability to upgrade into rapid transit, but lower frequency for a given demand due to larger vehicles, a need to introduce new facilities for a different technology in case of a new application, lower ability to branch out and hence requiring more transfers, and a longer implementation period. Figure 2.19 shows the picture of the LRT, and the capacity of LRT is given in Table 2.5.

2.4.3 Rapid Transit or Mass Rapid Transit System

These modes operate exclusively on category A ROW (i.e., a fully controlled ROW without grade crossings or any legal access by other vehicles or people) and have

FIGURE 2.19 Light Rail Transit

TABLE 2.5
Capacity of LRT

	Normal Load (5p/sq. m)	Crush Load (8p/sq. m)
Seats	65	65
Standees	160	256
Vehicle capacity	225	321
2-Vehicle train capacity	450	642
Line capacity (h–60 seconds) in pphpd	27,000	38,520

Source: Verma and Dhingra, (2001)

high speed, capacity, reliability, and safety. All existing rapid transit systems utilize guided technologies (rail or rubber tire) that permit operation of trains (high capacity) and automatic signal control (high safety). This class includes the following modes:

1. *Rubber-Tired Rapid Transit (RTR):* RTR (Figure 2.20) consists of moderately large vehicles (gross floor areas between 36–53 sq. m and 380–570 sq. ft) supported and guided by rubber tires, running on wooden, steel, or concrete surfaces in trains of five to nine cars.
2. *Rail Rapid Transit (RRT):* RRT typically consists of large four-axle rail vehicles (area up to 70 sq. m to 750 sq. ft) which operate in trains of up to ten cars on fully controlled (A) ROW which allows high speed, reliability, capacity, rapid boarding, and fail-safe operation (in case of drivers' error or disability, the train is stopped automatically) (Figure 2.21). The capacity of RRT is shown in Table 2.6.

FIGURE 2.20 Glimpse of an RTR

FIGURE 2.21 Glimpse of an RRT

TABLE 2.6
Capacity of RRT

	Normal Load	Crush Load
Seats	98	98
Standees	300	480
Vehicle capacity	398	578
3 Twin-vehicle capacity (6 vehicles)	1194	1734
Line capacity (h–90 seconds)	47,760	69,360

Source: Verma and Dhingra, (2001)

FIGURE 2.22 Regional Rail

TABLE 2.7
Capacity of RGR

	Normal Load	Crush Load
Seats	60	60
Standees	270	432
Vehicle capacity	330	492
Train capacity (6 vehicles)	2970	4428
Line capacity (h–3 minutes)	59,400	88,560

Source: Verma and Dhingra (2001)

3. *Regional Rail transit (RGR):* RGR, usually operated by railroads, has high standards of alignment geometry (Figure 2.22). It utilizes the largest vehicles of all transit systems (up to 80 sq. m to 860 sq. ft) which operate on trains, on longer routes, with fewer stations, at higher speeds than typical for RRT. Thus, RGR functionally represents a large scale RRT which serves most efficiently regional and longer urban trips. Capacity of RGR is shown in Table 2.7.

As per Verma and Dhingra (2001), the following are the capacity ranges, within which a particular mass transit system is suitable for operation in Indian cities:

Street transit system = Up to 12,000 pphpd

(Mini-bus, single-decker standard bus, double-decker, and articulated)

LRT1 = 12,000–36,000 pphpd

LRT2 = 36,000–50,000 pphpd

RRT (Metro) = 50,000–69,000 pphpd

RGR = 59,000–89,000 pphpd

TABLE 2.8
Recommendation of Mass Transit Systems for Indian Cities

Population (in Million)	Activity Structure[(2)]	Recommended Mass Transit System[(1)]		
		Circular City	Semicircular City	Linear City
13	PNU	RGR+LRT1/LRT 2+ST	RGR+LRT1/LRT 2+ST	RGR+RRT+LRT 1/ LRT2+ST
	PU	RGR+LRT1+ST	RGR+LRT 1/LRT 2+ST	RGR+RRT+LR T1/ LRT2+ST
	M	—	—	—
6	PNU	RRT+LRT 1/LRT2+ST	RRT+LRT 1+ST	RGR+LRT 1/LRT2+ST
	PU	RRT+LRT 1/LRT2+ST	RRT+LR T2+ST	RGR+LRT1/LRT2+ST
	M	—	—	—
3	PNU	LRT1+ST	LRT2+ST	RRT+LRT1/LRT2+ST
	PU	LRT1+ST	LRT2+ST	RRT+LRT1/LRT2+ST
	M	LRT1+ST	RRT+LRT1+ST	RRT+LRT1/LRT2+ST
1.5	PNU	ST	LRT1+ST	LRT1+ST
	PU	ST	LRT1+ST	LRT1+ST
	M	LRT1+ST	LRT1+ST	LRT1+ST
0.75	PNU	ST	ST	ST
	PU	ST	ST	ST
	M	ST	ST	ST
0.375	PNU	ST	ST	ST
	PU	ST	ST	ST
	M	ST	ST	ST

(1) RGR = regional rail transit, RRT = rail rapid transit, LRT1 = light rail transit1, LRT2 = light rail transit2, ST = street transit

(2) PNU = polynuclear nonuniform, PU = polynuclear uniform, M = mononuclear

Source: Verma and Dhingra, (2001)

The above ranges can be taken as the basis for determining the suitability of transit systems for Indian cities of different population sizes and forms based on the pphpd count on major corridors within the city. Further, Table 2.8 (Verma and Dhingra, 2001) shows the recommended combination of mass transit modes for Indian cities of different population size and structure.

2.4.4 Special Transit

Specialized transport systems are those which may have a role to play in a specific part of a city, without in any way forming a substantial part of the urban transport network. These are as follows:

1. *Magnetic Levitation:* In levitation systems, the coach is suspended in the air by magnetic levitation or by air cushion. The reason for the development

FIGURE 2.23 Monorails

of these technologies was to overcome the problems of vibration and resonance, which make it virtually impossible for normal trains to exceed 300 kmph. The air cushion system is still experimental and is not seriously considered. Until now, only small systems have been constructed at low capacities of 2000–5000 pphpd. Under Indian overloading patterns, the magnets may need to be augmented and may need remagnetizing more often.

2. *Monorails:* They have been designed for their low guideway cost, but have three main drawbacks:
   a. Low capacity (up to 15,000–20,000 pphpd)
   b. No form of emergency evacuation, so a low safety factor
   c. Complicated guide wheel system.

   Figure 2.23 shows a picture of monorails.

   Primarily since the monorail is designed for low capacities, and cannot meet the necessary demand levels in India, it has not been seriously considered, until now.
3. *Waterborne Transport:* Waterborne transport should be taken more seriously since most cities are either by the sea or on a river, and water transport may be able to take some of the load if traffic demand is across or along water.
4. *Automated Guided Transit (AGT):* AGT (Figure 2.24) modes are a low-capacity rail-based system of lightweight construction and totally automated. They consist of two groups: personal rapid transit (PRT), with small vehicles serving individual parties only, and group rapid transit (GRT), also known as people mover systems (PMS), with somewhat larger vehicles (15–50 spaces) designed mostly for short-haul medium capacity lines (Table 2.9). AGT is classified as special transit, together with other proposed and specialized modes. This class of transit contains both supported and suspended

FIGURE 2.24 Automated Guided Transit

types of technologies. These modes basically provide service in such areas as shopping centers, commercial areas, airports, etc. as an enhancement to the area's activities.

2.5 OPERATING ELEMENTS

The performance and efficiency of a transit system are determined by several key operational factors. These elements influence service quality, capacity, and overall effectiveness in meeting passenger demand. Understanding these factors helps in designing a well-functioning system that balances speed, accessibility, and resource use. Detailed description of the elements and formulations is provided in Vuchic (2005).

2.5.1 Transit Lines, Network, Stops, and Stations

A transit line is a predefined route that vehicles follow, connecting different locations. A network is formed when multiple lines intersect or connect, allowing passengers to transfer between routes and travel across a broader area. Well-planned networks improve accessibility and convenience by reducing travel time and enhancing connectivity.

Transit systems include stops and stations where passengers board and exit vehicles. Stops are typically used for buses and streetcars, while stations provide additional infrastructure such as platforms, ticketing systems, and waiting areas, often seen in RRT and bus rapid transit (BRT) systems. The spacing of stops and stations affects both travel time and accessibility—closer stops improve convenience, but fewer stops can result in faster service.

TABLE 2.9
System Characteristics of Some Selected Rail-Based Transit Systems

	Heavy Rail			Medium Capacity				
	EMU	Metro	Heavy Rail Recommended	LRT	Purpose Designed Lin. Mtr.	Mag. Lev.	Monorail	Sky Bus
Gauge (M)	1.667	1.667	1.667	1.435	1.435	1.435	N/A	1.435
Coaches/train	9	8	9	3	6	4	4	4
Length * width (m)	20* 3.66	20* 2.74	20* 2.74	29*2.5	16*2.65	12*2.3	10*2.5	8*2.8
Tare weight (T)	350	300	244	123	99	36	40	N/A
Payload/tare ratio	0.5	0.52	0.64	0.585	0.73	0.88	0.75	N/A
Axle load (T)	19	16.5	16	8	8	N/A	6	N/A
Acc./Dec/(m per second^2)	1.1/1.2	1.1/1.2	1.0/1.3	1.0/1.3	1.0/1.3	1.0/1.3	1.4/1.5	1.3
Max. speed (kmph)	80	80	80–120	80	80	80	80	100
Min. radius (m)	200	200	200	20	20	20	20	100
Max. gradient (per)	2	2	2	7	10	10	10	1.3
Power supply	25KV AC	750V DC	750V DC	750V DC	750V DC	1KV AC	750V DC	750V DC
Energy consumption (W hr./Ton. km)	40	52	40	34	40	35	40	N/A
Min. headway (Mts)	3	1.5	1.5	1	1	1	1.5	1

Source: Compiled from GOM (2000)

2.5.2 Vehicles, Transit Units, and Fleet Size

The vehicles used in transit systems vary widely, from standard buses and trams to high-capacity metro trains. A transit unit (TU) refers to a single operational entity, such as a bus, a tram, or a train composed of multiple connected cars. Fleet size represents the total number of vehicles available for service. Maintaining the right fleet size is crucial for balancing service reliability, passenger demand, and operational costs.

2.5.3 Passenger Flow and Volume

Passenger flow describes how individuals move through the system, including boarding, alighting, and transferring between routes. Passenger volume refers to the total number of people using the system over a certain period, such as an hour or a day. Managing passenger flow effectively is essential to prevent overcrowding, reduce delays, and maintain smooth operations, especially during peak travel hours.

2.5.4 Headway and Frequency

Headway is the time gap between two consecutive vehicles on the same route, while frequency is the number of trips operated per hour. Shorter headways and higher frequencies result in reduced waiting times and improved service quality, making transit more attractive to passengers. However, increasing frequency requires more vehicles and operational resources, making it important to find a balance between service efficiency and cost.

2.5.5 Capacity, Work, and Utilization

The capacity of a transit system is measured at different levels:

- Vehicle capacity refers to the number of passengers a single vehicle can accommodate.
- Line capacity is the maximum number of passengers a transit corridor can handle within a specific timeframe.
- Network capacity considers the overall capacity of the entire transit system across multiple lines and modes.

Work in transit operations is typically measured in terms of the total distance traveled by vehicles or the total seating capacity provided. Utilization indicates how effectively available capacity is used, measured by the proportion of actual passengers relative to total capacity. Higher utilization improves efficiency, but excessive crowding can reduce passenger comfort and service quality.

2.5.6 Travel Time and Speed

Travel time in a transit system includes the time spent in the vehicle, waiting at stops, and any additional delays due to congestion or transfers. Speed depends on factors such as stop spacing, ROW conditions, and priority measures like dedicated bus lanes or signal priority at intersections. Faster services attract more passengers, but if stops are too far apart, accessibility may decrease. A well-designed system finds the right balance between speed and convenience.

2.6 STATUS OF PUBLIC TRANSPORT WITH RESPECT TO COVERAGE, FLEET STRENGTH, AND UTILIZATION IN URBAN AREAS

2.6.1 Fleet Utilization (PER) and Fleet Strength

Fleet utilization is defined as the percentage of buses put on the road to the number of buses held. The overall fleet utilization of state transport undertakings (STUs) is given in Figure 2.25. It can be seen from the figure that the percentage of fleet utilization is decreasing from 2011 to 2020, whereas the number of STU buses is considerably increasing from 130,000 in 2011 to around 138,000 in 2019. After the year 2019, there was a slight dip in the number of buses due to the COVID-19 pandemic.

2.6.2 Public Transport with Respect to Coverage

From Table 2.10, it is clear that the public transport coverage of STU buses in terms of effective kilometers, dead kilometers, and gross kilometers is going on increasing from 2001 to 2019.

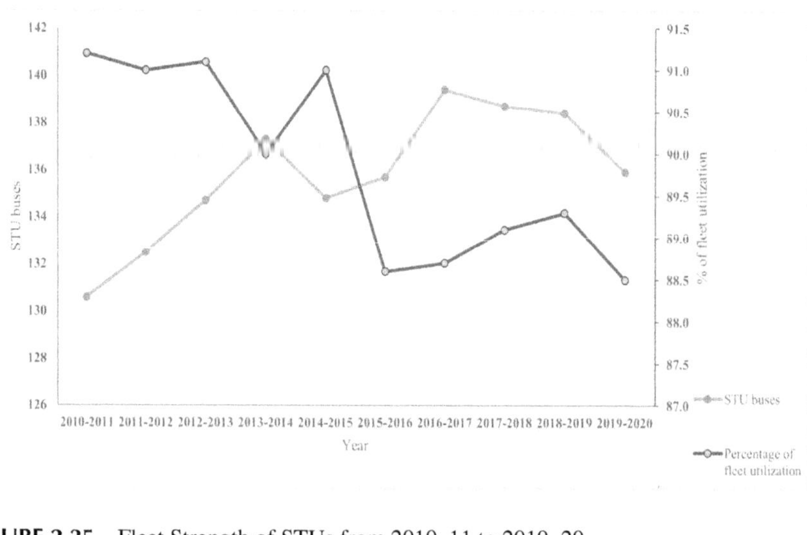

FIGURE 2.25 Fleet Strength of STUs from 2010–11 to 2019–20

Source: State Transport Undertakings (Central Institute of Road Transport)

TABLE 2.10
Public Transport with Respect to Coverage of STU Buses

Sl. No.	Year	Effective km (in Million)	Dead km (in Million)	Gross km (in Million)
1	2001–02	1250.644	32.132	1282.776
2	2002–03	1054.966	30.932	1085.898
3	2004–05	1182.724	31.286	1214.010
4	2005–06	1273.293	40.4	1313.693
5	2007–08	1306.280	50.074	1356.354
6	2008–09	1404.215	51.768	1455.983
7	2010–11	1618.323	59.796	1678.119
8	2019–20	14,785.861	252.403	15,038.264

Source: State Transport Undertakings (Central Institute of Road Transport)

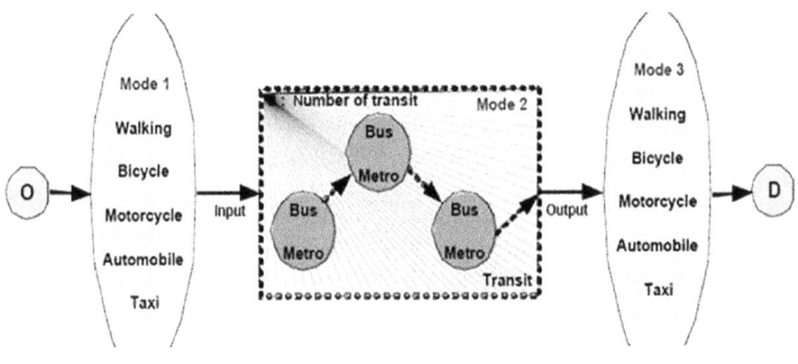

FIGURE 2.26 An Intermodal Trip

2.7 THE NEED FOR AN INTEGRATED TRANSPORT INFRASTRUCTURE APPROACH

Consider an intermodal trip between any given origin (O) and destination (D), as shown in Figure 2.26. The same can be done using various mode options and their corresponding infrastructure, in such a way that it optimizes the complete journey of travelers from origin to destination. While we plan the integrated transportation infrastructure to improve mobility and accessibility for all, it is important that we consider various scenarios consisting of combinations of different modes and infrastructure options and adopt the one that is most sustainable in terms of both mobility/accessibility and safety.

Cities like London, Zurich, Berlin, Paris, Munich, Hong-Kong, and Bogota provide good examples of well-integrated multimodal transport systems that provide seamless O–D connectivity through sustainable mode options.

TABLE 2.11
Details of the Two Corridors Approved by the Government of India for Bengaluru City

	Phase 2 Nagasandra to Madavara		Phase 2A Central Silk Board to Kadubeesanahalli		Phase 2B Kasturi Nagar to Kempapura	
Corridors	**Length (km)**	**Stations**	**Length (km)**	**Stations**	**Length (km)**	**Stations**
Elevated	29.66	25	19.75	13	38.44	17
Underground	13.88	12	—	—	—	—
Surface	—	—	—	—	—	—
Total	43.54	37	19.75	13	38.44	17

Source: BMRCL

2.7.1 Cost of the Metro Project and Corridors

Taking the Bengaluru Metro as an example, as far as rail systems are considered, it is not always possible in present urban conditions to go for any particular (surface, underground, or elevated) type of construction; hence, a network should not be thought of as solely elevated or surface or underground, but rather the three should complement each other in a network, and fit in with the requirement of each individual section. In addition to the existing network of metro in Bengaluru, Table 2.11 shows the details of the ongoing project corridors approved by the Government of India for Bengaluru City.

2.8 COMPARATIVE ANALYSES

The procedure for the comparative analysis and selection of transportation modes follows these major steps:

Step 1: Based on urban transportation policy, develop goals for the transit system.
Step 2: Define conditions for the area to be served.
Step 3: Utilizing results from preceding steps, define specific requirements and standards for the planned system.
Step 4: Select ROW type for candidate modes.
Step 5: Select technologies and type of operation for candidate modes.
Step 6: Develop functional designs for candidate modes.
Step 7: Evaluate candidate modes.
Step 8: Compare evaluation results (based on capacity, cost, environmental impact, etc.) and select the optimal mode.

From the results obtained (Table 2.8), it can be seen that in cities with populations up to 0.75 million (also circular PNU and PU cities of population 1.5 million), the

TABLE 2.12
Maximum Peak Hour Public Transport Demand in pphpd for Indian Cities

Population (in Million)	Public Transport Share Factor	Activity Structure	Maximum Peak Hour pphpd Circular City	Semicircular City	Linear City
13	0.84	PNU	168,840	252,000	388,080
		PU	120,960	297,360	383,040
		M	—	—	—
6	0.74	PNU	66,600	100,788	155,400
		PU	57,720	119,880	150,960
		M	—	—	—
3	0.64	PNU	27,264	38,400	61,440
		PU	24,192	49,920	62,208
		M	33,408	65,280	65,280
1.5	0.58	PNU	10,440	16,356	24,360
		PU	9744	19,488	24,360
		M	13,224	26,796	26,796
0.75	0.55	PNU	3960	5940	8910
		PU	3300	6930	8910
		M	4620	9900	9900
0.375	0.53	PNU	477	859	1272
		PU	636	954	1272
		M	700	1272	1272

street transit can fully cater to the peak hour passenger travel demand, as the maximum peak hour public transport demand is within the capacity of the street transit (Table 2.12). However, for cities with populations above these values, a rail-based system along with a street transit system is required to cater to increased passenger travel demand. Also, it can be seen from Table 2.8 that cities with population greater than 6 million probably require the highest order mass transit systems like RGR to meet the heavy passenger travel demands (Table 2.12) on major corridors within the city.

Finally, Table 2.13 shows in tabular form some of the more frequently encountered relationships involved in translating system goals into design objectives and, in turn, into design methods and technology attributes. For example, to meet the goal of minimizing construction costs, the designer must choose the objective of maximizing the use of shared facilities, seeking to run on high occupancy vehicle (HOV) lanes and transit ways open to carpools and vanpools. In this case, BRT would be the favored technology, being capable of operation in all types of HOV configurations, LRT would be the second choice, being appropriate for limited running in arterial HOV lanes only, and the last choice would be RRT and AGT with their requirements for exclusive guide ways.

The authors feel that the chapter would strongly help planners in selecting a suitable mass transit system for Indian cities of different sizes and forms.

TABLE 2.13
Some Relationships between Design Goals and Technology Attributes

Goal/Design Objective	Design Method	Technology Suitability Rank			
		RRT	AGT	LRT	BRT
Maximize ridership:					
Locate station within easy walk of	Locate system underground to allow unobtrusive/no disruptive high-capacity entry into high-density areas	1	2	2	3
Many major centers	Locate system in surface streets/malls of major centers, with first floor-level stops	2	2	1	1
	Use high line mileage and many stations system wide	1	1	1	1
Provide high-frequency service	Use short trains or single-vehicle trains with short headed ways	3	1	2	1
Maximize scheduled speed	Provide grade separation and high-speed alignment for entire system	1	2	3	4
	Provide skip-stop and express service	2	2	2	1
Reorganize transit service system wide	Remove radial bus service; provide focus to reorient bus route into community/cross-town operation	1	2	3	4
Maximize development impact:					
Stress accessibility and permanence	Use fixed guide way with substantial stations centers to areas of potential development/redevelopment	1	2	3	4
Minimize construction cost:					
Use of existing ROW to avoid underground/elevated construction	Use freeway medians, railroad/powerline ROW, though these may be distant from activity centers	1	1	1	1
	In lower-density areas let system run on streets/highways mixed with other traffic	3	3	2	1
Maximize use of shared facilities	Run on HOV lanes and other facilities open to carpools/vanpools	2	2	2	1
Reduce total construction required	Reduce system mileage, number of stations	1	1	1	1
	Use shorter, simpler stations, low platforms.	2	2	1	1
	Use smaller horizontal and vertical clearances, lighter structures	2	1	2	2
Reduce system complexity	Eliminate power distribution and control system	3	3	2	1
Minimize operating cost:					
Reduce operating personnel	Use long trains to reduce personnel/passenger ratio	1	2	2	3

TABLE 2.13 (*Continued*)
Some Relationships between Design Goals and Technology Attributes

Goal/Design Objective	Design Method	Technology Suitability Rank			
		RRT	AGT	LRT	BRT
	Use a more complex system affording greater automation	2	1	3	4
	Use short trains in off-peak	3	1	2	1
Reduce maintenance personnel	Use simpler system with less electronics and hardware	3	4	2	1
Maximize public support:					
Provide service to widest possible area	Use low cost/mile systems, maximum use of at-grade, nonexclusive ROW	4	3	2	1
File predispositions of public	Use rail/fixed guide way systems; avoid bus systems	1	1	1	2

2.9 A LOOK AHEAD IN PUBLIC TRANSPORTATION

In the dispersed regional city of the future, public transportation will be needed to carry out five essential functions:

1. To guarantee city-wise mobility for the growing number of people who are nondrivers by choice or necessity.
2. To supply the exclusive means of travel in high-density areas where cars are prohibited.
3. To complement the services rendered by the automobile on trips that requires both methods.
4. To provide local extensions of the intercity and global public transportation networks.
5. To help create a more satisfying, manageable, pollution-free, and sustainable urban environment that maximizes the ability to move while minimizing the necessity for movement. Harman (1998) outlined the major advances anticipated in the different forms of public transportation. These included the following:

   Buses: There will likely be production of a bus that has two methods of propulsion in one coach, a diesel engine for use on suburban streets and electric power for city streets and in tunnels. Some innovations can be expected in alternative forms of access to the vehicle, in particular for the elderly and handicapped. Although some concepts are currently being developed that would make the bus–road interface much more intelligent in terms of navigation and vehicle control, such programs are not likely to see widespread implementation in the early part of the twenty-first century.

Heavy and Light Rail: The major source of innovation will likely come in the means of providing propulsion and in automatic system control.

Commuter and Intercity Rail: There were possible major advances that could be anticipated in the application of linear induction motor, magnetic levitation technology, and other forms of providing high-speed passenger transportation.

Automated Guide Way: The technology for implementing automated guide way transit, group rapid transit, or personal rapid transit systems is already available. The key issue is now to find the appropriate applications.

Harman then suggested that perhaps the greatest technological innovations will come in the areas of transit user information systems, automatic vehicle monitoring, and in revisiting already tested applications such as high-occupancy vehicle facilities.

SUMMARY

This chapter mainly deals with the different transport modes and their characteristics which are mainly classified into three modes, that is, personal, intermediate, and public transport modes. The transit modes are classified into street, semi-rapid, and special transit. These transit modes consist of different types and their characteristics. This chapter also covers detailed capacity assessment of some selected technologies, and integrated transport infrastructure approach, besides information about the various transport modes and their role with specified figures and graphs.

3 Current Scenario of Public Mass Transport (PMT) in Developing Countries

3.1 CURRENT PERCEPTION AND USAGE OF PMT IN THE DEVELOPED AND DEVELOPING NATIONS

Public transport services play a vital role in the overall development of a country, especially in reducing emissions per capita. The living standard in many developing countries is changing. While developing cities are showing improvements in economic growth in terms of GDP, middle class people are no longer attached to the particular transport services; higher income society is going to comfortable travel times and higher expectations in terms of better mobility and better quality of services. Public transport usage in these economies is often necessity-driven, with low- and middle-income groups relying on it due to affordability. However, aspirations for private vehicle ownership among the middle class have led to a gradual decline in public transport use for certain demographics. Conversely, in developed economies, public transport remains a competitive alternative to private vehicle ownership due to its reliability, convenience, and environmental benefits, appealing to a broader demographic base. Policies such as subsidies, integrated ticketing systems, and congestion pricing ensure public transport remains a preferred option, even for high-income groups. Usage is also supported by extensive networks that allow for seamless travel within and across urban regions.

Public transport systems are known to be sustainable modes in terms of space and energy efficiency as well as environmental and social benefits. A good, networked public transportation system has some of the required traits, such as time-bound schedules, reliable services, comfort, competitive travel times, and affordable prices, for providing sustainable transport services and commuter satisfaction. In developing economies, these traits are often not fully realized due to overcrowding, service inconsistencies, and safety concerns. These issues limit usage, especially among those with alternatives, despite the critical role that public transport plays in ensuring mobility for the majority of the population. Passenger transportation has an impact on all aspects of mobility and is an important part of overall economic development. Improving the performance of public transport undertakings is becoming increasingly critical due to the paucity of public funds, increased demand on transport services, and expanding social needs. Recently, the performance measurement

DOI: 10.1201/ 9781003740247-3

and evaluation systems have been gaining importance (Kittleson Associates et al., 2003; Sulek and Lind, 2000).

Increased urbanization has increased the number of passenger vehicles in the cities in developing countries such as India. This trend has two significant effects: a reduction in the share of public transport usage among the middle and upper classes and worsening congestion, which further diminishes the efficiency of public transport systems. To address these challenges, strategic interventions such as investment in infrastructure, improved service delivery, and policies encouraging public transport adoption are needed. The Road Transport Corporations Act came into effect in India in 1950 and led to various state governments setting up respective State Road Transport Corporations with an objective of providing affordable transport services within the state as well as across states. Over the years, most of these corporations have become loss-making entities. The trade-off between commercial objectives and social responsibility goals of these state-owned corporations became an issue of major concern. In contrast, developed economies often sustain high public transport usage through consistent funding, regular modernization, and governance frameworks that prioritize reliability, comfort, and integration with other modes of transport.

3.2 CASE STUDY

Balancing between social and commercial objectives requires innovative solutions and the adoption of best practices from successful models in developed economies, where public transport operates as a financially sustainable and socially inclusive system. In the work of Verma et al. (2013), the SERVQUAL framework, which is a multi-item instrument used for measuring service quality, is used to measure gaps between commuters' expectations of urban bus transport services and the actual service quality provided by the Bangalore Metropolitan Transport Corporation (BMTC).

3.2.1 Bengaluru

BMTC is taken as the case study for applying the proposed approach. BMTC operates about 6000 routes in Bengaluru city and has a fleet of more than 6500 buses with a considerable number of luxury (Volvo and Mercedes-Benz) buses to its credit. It carries about 3.5 million passengers every day and has a very efficient bus staff ratio of around 6. While the services of BMTC are very competitive among any other urban public bus transport organization in the country, they are still to reach the desired level of passenger mode share that will enable them to provide sustainable mobility solutions to the commuters in Bengaluru city. Also, while BMTC is currently modernizing the bus fleet, to improve its services and mode share, through funding under Jawaharlal Nehru National Urban Renewal Mission (JNNURM) scheme, there is still very little focus on improving overall service quality to improve commuter patronage. The outcome of the study may help BMTC to bridge the gap between commuters' expectations and actual service quality and to work toward developing sustainable services.

3.2.2 Working Process

The objective of the study is to measure gaps between commuters' expectations of urban bus transport services and the actual service quality provided. The SERVQUAL framework is used to measure the gap between expectations of public transport services of customers (commuters) and the actual service quality provided. SERVQUAL is a multi-item instrument for measuring service quality. This instrument was first developed by Parasuraman et al. (1985) through an exploratory study of marketing academics. The outcome was a 22-item scale that has received widespread application in the research of service quality. The 22 items are essentially framed around the following five dimensions of service quality:

1. *Tangible:* Physical facilities and equipment.
2. *Reliability:* Ability to perform the promised service dependably and accurately.
3. *Responsiveness:* Willingness to help customers and provide prompt service.
4. *Assurance:* Knowledge and courtesy of employees and their ability to inspire trust and confidence.
5. *Empathy:* Caring and individual attention the firm provides to customers.

Using the above five dimensions of service quality, parameters related to the case study were identified under each of the five dimensions. Accordingly, a survey questionnaire was developed to obtain the required data for analysis. Finally, statistical analysis using a paired *t*-test was done to identify, measure, and interpret the gaps between perception and expectation. The next section discusses the case study application.

3.2.3 Data Collection and Analysis

3.2.3.1 Sampling

The main intent of this study is to measure gaps between commuters' expectations of urban bus transport services and the actual service quality provided. The empirical data were collected using a questionnaire. Overall, 196 samples of bus commuters were selected from different regions of Bengaluru, using snowball sampling technique. The questionnaires were distributed personally and by using social media networks like Facebook. An online version of the questionnaire was also distributed using Google Docs to the potential respondents. Out of 196 responses, 186 were usable, as 10 questionnaires received were partially filled and could not be used for analysis. This shows 94.87% response rate. Thus, the actual sample size used for analysis was 186.

3.2.3.2 Variable Measurement

A total of 50 questions were used to measure the service quality of bus services in Bengaluru based on five dimensions defined by Parasuraman et al. (1985) as mentioned above. All questions were designed as close-ended questions on a 7-point Likert scale varying from strongly disagree (1) to strongly agree (7).

3.2.3.3 Sample Profile

The demographic and travel profile of the respondents is shown in Tables 3.1–3.3 which includes gender, age group, marital status, education profile, income range per month, occupation of respondents, and various types of buses used by the respondents and frequency of bus travel on an average.

3.2.4 Survey Results

3.2.4.1 Demographic Profile of Respondents

There is a notable difference in the genders using urban bus services from the respondents. Sixty-three percent of the total respondents were male commuters and 36% were female; 67.74% of the users were in the age group of 20–25, which shows that the commuters (among the respondents) are mostly the young generation, that is, the students and the neo-working class.

TABLE 3.1
Demographic Profile of the Respondents

Characteristics		No. of Respondents	Percentage of Respondents
Gender	Male	115	62
	Female	71	38
Age	Below 20	12	6
	20–25	123	66
	25–30	34	18
	35–45	6	3
	45 and above	11	6
Marital status	Married	35	19
	Unmarried	151	81
Education	Undergraduate	26	14
	Graduate	68	37
	Postgraduate	92	49
Income range/month	Below 5000	52	28
	5000–10,000	30	16
	10,000–20,000	48	26
	20,000 and above	56	30
Occupation	Student	125	67
	Government employee	4	2
	Own business	7	4
	Private company employee	38	20
	Others	12	6

TABLE 3.2
Types of Buses Used by the Respondents

Bus Type	No. of Respondents	Percentage of Respondents
Vajra	46	25
BIG 10	5	3
Suvarna	12	6
Ordinary	101	54
Pushpak	10	5
Others	12	6

TABLE 3.3
Frequency of Bus Travel

Frequency of Bus Travel	No. of Respondents	Percentage of Respondents
Daily	58	31
Occasionally once in a while	72	39
Rarely	56	30
Never	0	0

3.2.4.2 Service Quality Gaps

The questionnaire which was developed consisted of two sets of items with each set having 25 questions. The first set was used to measure the perceived LOSs quality (P). The same set of questions were reframed/rephrased to inquire about the users' expectations of the service quality (E). The questions under each dimension were rated using a 7-point Likert scale.

A gap analysis was conducted on each dimension of the SERVQUAL (Parasuraman et al., 1985) taking averages of individual items under each dimension and subtracting the expectations from the perceptions (P – E). The average of the perceptions scores and the expectations score results are tabulated in Table 3.4. The comparison of average of each of these dimensional scores is shown in Figure 3.1, which clearly shows that the perceptions of the commuters are below the expectations of the services provided by the BMTC in almost all the dimensions, and the gap is much wider in case of the empathy dimension.

3.2.4.2.1 Interpretation of the Gap Analysis

The following are three possible outcomes and their interpretation, for the gap analysis done in Table 3.4:

TABLE 3.4
Comparison of Perceptions and Expectations of Respondents

Dimensions	Average Perception Score (P)	Average Expectation Score (E)	Difference (P – E)
Reliability Dimension			
Bus information schedule and route maps are available and reliable	3.806452	5.451613	−1.64516
Buses are available on time during peak hours	3.704301	5.715054	−2.01075
Buses are available to every area in the city	3.994624	5.682796	−1.68817
Frequency of buses is very high on every route	3.456989	5.526882	−2.06989
Computerized ticketing system leaves little scope for cheating and bribing	5.021505	5.682796	−0.66129
Tangible Dimension			
Buses are clean and well maintained	4.064516	5.827957	−1.76344
Buses are a safe mode of transport	5.016129	5.887097	−0.87097
Buses are the best mode for advertising and campaigning	4.478495	4.930108	−0.45161
Eco-friendly buses are used	4.016129	5.822581	−1.80645
Bus stops are well maintained	3.301075	5.860215	−2.55914
Responsiveness Dimension			
Bus tickets are affordable, and buses are a real value for money	4.634409	5.763441	−1.12903
Bus routes are not lengthy	4.021505	5.306452	−1.28495
Bus stops are conveniently located	4.489247	5.548387	−1.05914
Response time to resolve complaints is very low	4.344086	5.365591	−1.02151
Bus information is easily available through calls, SMS, and on the internet	3.844086	5.811828	−1.96774
Assurance Dimension			
Drivers and conductors are courteous	3.860215	5.715054	−1.85484
There are a lot of safety measures against crime on buses	3.913978	5.741935	−1.82796
There is very little accident damage caused by buses	4.086022	5.682796	−1.59677
Drivers are well trained, and safety measures are taken care of	4.225806	5.88172	−1.65591
Fire and emergency exits are available on all buses	4.268817	5.865591	−1.59677
Empathy Dimension			
Bus is user friendly for handicapped	3.543011	5.973118	−2.43011
Seats are available on every bus	3.11828	5.44086	−2.32258
Buses are safe for young mothers	3.403226	5.860215	−2.45699
There is first aid available on every bus	3.096774	5.827957	−2.73118
Destination display systems are useful for visually impaired and aged	4.33871	5.930108	−1.5914

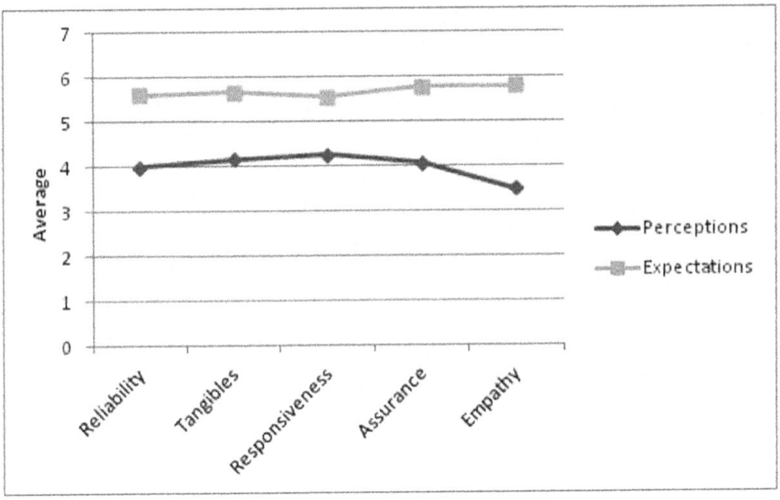

FIGURE 3.1 SERVQUAL Dimension Gaps Between Expectations and Perceptions

P – E >0, this implies that the users are satisfied with the LOS being provided.
P – E = 0, this implies that the satisfactory levels are just being touched or reached.
P – E < 0, this implies that the satisfaction levels are low.

An average of the perception and expectation scores of each of the dimensions in the questionnaire is taken and the respective difference is sought out in Table 3.4. The differences can be positive or negative value. A positive value indicates that the perceptions of the users have exceeded their expectations and have resulted in a satisfied customer, whereas a negative value shows that the expectations of the users have not been met by the service provider, and there is a scope for improvement in services.

Quite clearly, all the values in Table 3.4 are negative, that is, the service quality at present levels is not satisfactory for the commuters, which implies that there is a gap in the service quality offered and the expectations of the users. Analysis indicates that the bus services provided within the Bengaluru region have to improve in all the five dimensions of reliability, tangibility, responsiveness, assurance, and empathy dimensions as all the dimensions have negative scores. Commuters indicate the biggest overall service quality gap in the empathy dimension where commuters are asking for safety for young mothers, seat availability on the buses, provision for the disabled, availability of first aid and destination display systems for the visually impaired and aged. Each of the five dimensions is discussed in detail below.

3.2.4.3 Reliability Dimension

The reliability dimension elaborates the ability of the BMTC buses to perform the promised service dependably and accurately. This dimension elaborates the availability of service during peak traffic hours, reliability of public transport in all areas of the city as well as the trustworthiness of the computerized ticketing system which has been recently introduced. The negative scores on all these areas of focus show that the expectations are higher than perceptions and that the BMTC bus services should improve in all the areas of focus such as timely availability of buses during peak traffic hours, providing reliable transport to all areas of the city, and providing more buses for the convenience of commuters.

3.2.4.4 Tangible Dimension

The tangible dimension expounds the physical facilities and equipment of the BMTC buses in terms of safety, cleanliness, and maintenance of the buses as well as the bus stops. All the values in the gap analysis show negative values indicating that the BMTC bus service scores low again on the tangible dimension. Commuters' expectations are not met by the BMTC services; buses and bus stops are neither clean nor well maintained; and the safety of the commuters is not taken care of as expected by the public.

3.2.4.5 Responsiveness Dimension

The responsiveness dimension elaborates on the willingness of service providers to help customers and provide prompt services. The gap analysis shows the values are negative, and the expectations are not met in this case either. Overall, it shows that bus stops may not be conveniently located, bus routes are lengthy, and the response time to resolve complaints is high according to the perceptions of the commuters.

3.2.4.6 Assurance Dimension

This dimension accesses the knowledge and courtesy of employees and their ability to inspire trust and confidence in their services. The gap analysis shows all the values on a negative scale which implies that the overall BMTC service quality in the assurance dimension falls below the expectations of the commuters. Also, the wide gap with respect to all parameters of this dimension (refer to Table 3.4) shows that this dimension fares second worst after the empathy dimension. The overall interpretation of this gap will be that drivers and conductors are not well trained to behave courteously with the public, fire and emergency exits are not available on all buses, and there are instances of theft and crime that take place on the buses.

3.2.4.7 Empathy Dimension

The empathy dimension expounds on the care and individual attention the firm provides to customers by improvising and adding new features to the services which are user friendly to the differently abled, young mothers, and the safety of the passengers/commuters.

3.2.5 *t*-Test Analysis

To statistically validate the results, a paired sample *t*-test was conducted for all the dimensions to evaluate the statistical significance of gap between expectations, and perceptions and the results are shown in Table 3.5. The null hypothesis for each dimension is defined as follows:

TABLE 3.5
***t*-Test**

	Paired Differences						
Dimensions	Mean of P – E)	Standard Deviation	Standard Error	95% Confidence Interval of the Differences		t-value	Degree of Freedom (df)
				Lower	Upper		
Reliability	1.615	2.022	0.148	1.324	1.905	10.892	185
Tangibles	1.490	1.956	0.143	1.209	1.771	10.388	185
Responsiveness	1.292	1.998	0.146	1.005	1.579	8.818	185
Assurance	1.706	2.198	0.161	1.390	2.022	10.586	185
Empathy	2.306	2.259	0.165	2.631	2.631	13.921	185

H_0: There is no significant difference in commuters' perception and expectation of BMTC bus service quality in Bengaluru. As the sample size is large, the *t*-value corresponds to probabilities for standard normal distribution. The *t*-values for all five dimensions, that is, tangibility, reliability, responsiveness, assurance, and empathy dimensions, were found to be significantly high as shown in Table 3.5. At 5% level of significance, all the values above 1.96 in the *t*-table are considered as not satisfactory. Hence, there is not enough evidence to accept the null hypothesis. The results presented in Table 3.5 show that the *t*-values fall out of the upper limit considering 95% confidence limits and this implies that the expectations of the commuters fall above the services that are offered by the BMTC.

The overall results of this study also explain some of the reasons for the unsustainable trend in Indian cities, including Bengaluru, of exponential growth of car ownership and its usage and the corresponding transport externalities resulting from this trend. Due to overall higher levels of expectations than perceptions from bus transport and particularly a wider gap in the empathy dimension, commuters shift to cars as soon as they are able to afford it. In addition, Indian cities have higher levels of two-wheeler usage, which is cheaper and more flexible to use than buses.

This case study of BMTC from Verma et al. (2013) demonstrates an effective methodology for measuring current perception and service quality of public bus services.

3.3 PLANNING AND OPERATIONAL STRATEGIES FOLLOWED IN INDIA AND IN OTHER DEVELOPING COUNTRIES

In most large cities in developing countries, buses have historically been the primary mode of public transport, carrying a significant share of urban travelers.

While buses still serve as a crucial backbone of urban mobility in many areas, the public transport landscape has started to shift with the introduction and expansion of metro systems in several cities. Despite this progress, transit bus companies in these countries continue to face significant challenges. Transit bus companies in these countries are often cash strapped. In many cases, the operating cost per bus kilometer exceeds revenues and bus fares are often kept low irrespective of the cost of providing service. Many cities are dominated by old and fuel-intensive buses with high operating costs. Transit systems are also often plagued by overcrowded and undependable service, congested roadways, and chaotic operating environments.

Over the years, there has been significant progress in public transport systems in some developing cities with the introduction and expansion of metro rail networks. Cities like Delhi, Bengaluru, Jakarta, and Cairo have witnessed a shift in public transport usage patterns as metro systems provide faster, more reliable, and less congested alternatives to traditional bus services. This development has led to buses increasingly being used as feeder systems to metro stations, creating opportunities to integrate various modes of public transport into cohesive networks. Across the board, city officials in developing countries are under strong pressure to improve efficiency and enhance the attractiveness of bus transportation. In this context, upgrading bus fleets with cleaner and more energy-efficient vehicles, introducing smart ticketing systems, and improving the overall reliability and accessibility of bus services are critical steps. Additionally, integrating bus and metro services through well-planned multimodal hubs has emerged as a key strategy to create efficient and user-friendly public transport systems.

To understand the process of planning and operational practices of a developing country like India, Bengaluru, one of the cities of India, is taken as an example for the case study.

3.3.1 Transportation in Bengaluru

The Unified Metropolitan Authority (UMTA) is a framework designed to improve coordination and planning of urban transport systems in metropolitan areas. It integrates various stakeholders, such as municipal corporations, bus operators, metro authorities, and traffic management agencies, under a single authority. UMTA focuses on creating comprehensive mobility plans, streamlining policies, and optimizing resource allocation to enhance public transport systems and reduce traffic congestion. By promoting seamless integration between transport modes and adopting data-driven planning, UMTA aims to make urban transport systems more efficient, sustainable, and commuter-friendly. Several cities in India are in the process of establishing UMTAs to address the challenges of fragmented transport management and improve urban mobility.

BMTC continues to be a major provider of public transport in the city, although the landscape of public transport in Bengaluru has evolved significantly with the introduction of the Namma Metro system. Bengaluru has one of the most extensive networks of public bus transport in India. BMTC does not use private contractors currently and has no plans to use them in the future. It has around 50 operational depots spread throughout their region so there are minimal dead kilometers in getting to and from the bus starting and terminal points. BMTC owns its bus fleet. As per the data of 2019–20, the BMTC had around 140 lakh dead kilometers, which is approximately 3.3% of the gross kilometers travelled. BMTC even has its own bus body building centers. The Shantinagar workshop performs body building and reconditioning of assemblies. Its innovations in body building have improved aesthetics and resulted in fuel savings with better KMPL (kilometer per liter) and EPKM (earning per kilometer).

BMTC has a traffic department whose primary function is to plan and implement public transportation services. This department is involved in the assessment of demand for public transport, designing routes, route surveys, fare fixations, and fixation of bus stops among various other things. However, BMTC, just like most public transport corporations in India, does not have a dedicated transportation planning and modeling divisions. Most of the decisions are taken via demand-response techniques. Travel demand analysis is seldom used for planning present or future operations. Government of India changed this scenario by the creation of UMTA that includes all major stakeholders.

Bengaluru has one of the youngest fleets for a city transport corporation among other major transportation corporations with an average age of 6.9 years. However, around 17% of the existing buses are classified as overaged. BMTC operates many new Volvo buses which lead the industry in terms of vehicle design and comfort. The Vajra buses are tracked using GPS, and real-time arrival and departure information is displayed at the bus stops. Other than these services, about 4000 BMTC buses have GPS fitted into them, and BMTC has also set up a department which looks into collection, analysis, interpretation, and dissemination of data to help in the decision-making process. Most of the buses run on either diesel or petrol. BMTC has built ten Traffic and Transit Management Centers (TTMCs) as part of the development of urban infrastructure stated in the Comprehensive Traffic and Transportation Plan (CTTP) for Bengaluru and vision plan under JNNURM. These TTMCs are conceived as transport infrastructure for the urban renewal project aimed at providing one-stop travel to commuters. The TTMCs comprise state-of-the-art bus terminals, maintenance facilities, public amenities, information systems, safety and security, park and ride facilities, and they provide intermodal connectivity. TTMCs integrate transport facilities that cater to diverse user needs while ensuring efficient and smooth traffic flow to and from terminals without disrupting main roads. They aim to reduce congestion, minimize conflicts between different traffic types, and optimize passenger and vehicle processing times. Additionally, the TTMCs promote the use of public transport through the provision of park-and-ride facilities, enhancing first-mile connectivity and encouraging sustainable urban mobility.

3.3.2 Published Schedules and Route Searches

A major shortcoming in many Indian public transport systems has historically been the lack of accessible and reliable information on bus routes and schedules. Passengers, especially those new to a city or unfamiliar with public transport, often struggled due to the absence of route maps, stop details, and timings at bus stops or online. However, recent advancements in technology have significantly bridged these gaps, transforming public transport accessibility.

Cities like Bengaluru have made notable progress in this area. BMTC has moved beyond basic route search tools and now incorporates real-time data systems powered by GPS technology through its mobile app "Namma BMTC" and other platforms. Passengers can track buses live, view estimated arrival times, and get notifications about delays or route changes. This is complemented by dynamic display boards at major bus stops, which show real-time bus arrival updates, ensuring seamless journey planning for commuters. Similar advancements are seen in other Indian cities. These initiatives are helping public transport systems become more user centric.

Furthermore, the integration of public transport data with global platforms like Google Maps and local transport apps has made route and schedule information more accessible to a wider audience. Passengers can input their starting point and destination to receive optimized travel plans, including bus timings, estimated travel duration, and transfer points. These advancements have transformed how users interact with public transport, making it more efficient, user-friendly, and reliable.

3.3.3 BMTC Bus System Performance

BMTC is one of the few public transport corporations in the world that is profitable. They have been consistently increasing the number of buses and kilometers run every year and also the profit every year since 2005. Table 3.6 shows the detailed statistics from 2004 to 2022.

TABLE 3.6
BMTC System Performance Indicators, 2004–09

Parameter	2016–17	2017–18	2018–19	2019–20	2020–21	2021–22
Fleet size	6165	6677	6521	6690	6521	6799
Bus kilometers (in lakhs)	4205.2	4164.53	4152.85	4065.51	2147.48	2484.23
Operational costs (INR)	236,701.42	244,461.00	264,278.29	266,931.0	201,725.88	229,336.91
Fare revenue (INR)	177,047.15	176,470.60	183,884.04	180,700.00	69,976.42	92,249.14
Other revenue (INR)	33,563.26	46,228.91	45,445.51	31,297.00	115,329.48	119,263.23
Passengers carried per day (in lakhs)	45.34	44.37	35.00	33.10	12.79	15.97
Passengers/bus	879	804	624	597	352	402

3.3.4 Operational Statistics

Annual bus kilometers and ridership for BMTC experienced a decline compared to 2018–19 levels, largely due to the impact of COVID-19 and improved operations of metro system, particularly between 2019 and 2021. However, in the years following the pandemic, BMTC has demonstrated a steady recovery, with consistent increases in both the total kilometers travelled by buses and annual ridership figures. This upward trend reflects a gradual return to normalcy and growing reliance on public transport among commuters. These fluctuations in ridership have impacted fleet size as well. The dip in the numbers of 2023–24, as can be observed from the figures, may be due to the limited availability of data (Figures 3.2–3.4).

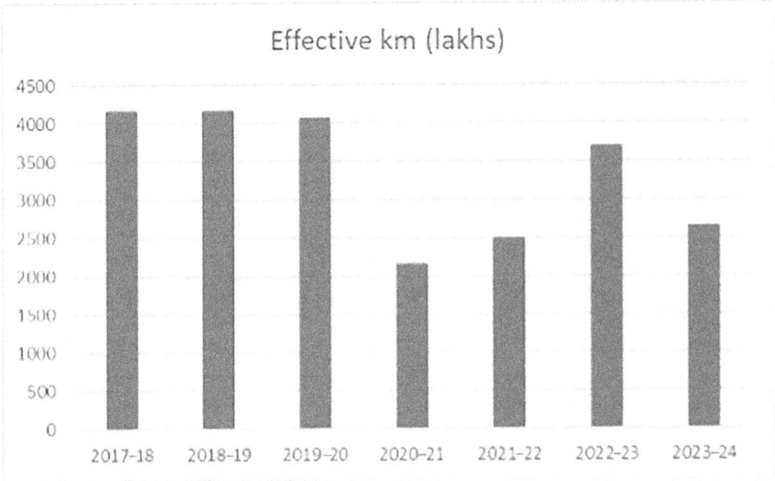

FIGURE 3.2 BMTC Annual Bus Kilometers Operated, 2017–24

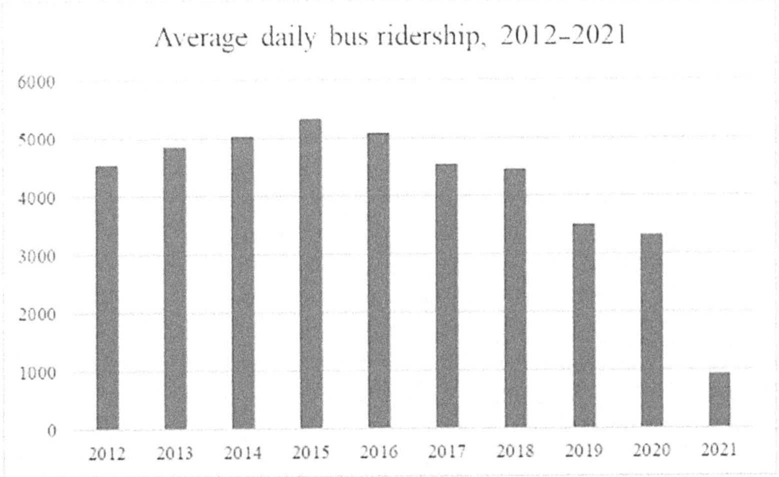

FIGURE 3.3 BMTC Average daily bus ridership, 2012–21

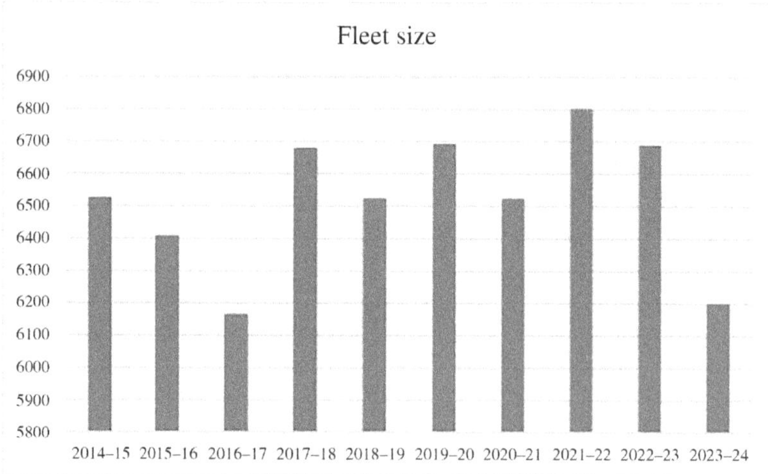

FIGURE 3.4 BMTC Bus Service Fleet Size, 2014–24

3.4 ISSUES OF COORDINATION WITH RESPECT TO FIRST-MILE AND LAST-MILE CONNECTIVITY AND LOSs

Infrastructure is the basic physical and organizational structure needed for the operation of services and facilities of a system. For a transportation system, the infrastructure will mean roads and highway networks, mass transit system, railways, walking and cycling facilities, waterway network and facilities, airways, etc. Here, a service is nothing, but an intangible commodity provided to facilitate the movement of people and goods from one place to the other using the available transportation infrastructure. The presence of integrated transportation infrastructure is a key requirement to ensure provision of effective, smooth, reliable, and safe services.

Decisions about transport systems, the form of urban development, and how land is used impact one other. It is therefore important to take a holistic or, in other words, a systems approach to transportation infrastructure planning wherein we treat all individual transportation facilities or projects as part of a single interrelated system and thus provide for both safe and fast movement of people and goods from point A to point B in a given physical space. The present state of transport externalities that Indian cities are experiencing today is the result of negligence toward the need for such an integrated or holistic approach. Various projects pertaining to individual modes or infrastructure categories are planned and executed in isolation and without understanding their impact over the whole network and system. This is due to the fact that each item is planned and implemented by a different agency. Thus, there is not enough operational coordination among different regions in implementing UMTA. For example, often flyovers are planned over junctions to improve local conditions but without understanding its impact over the whole citywide network flows. Further alternative analysis with various feasible options is seldom done to ensure provision of the best solution within the given practical constraints. Similarly, master plans are never prepared for integrated public transport systems, rather individual and separate studies are done for introducing, say, metro rail or BRTS system without effectively treating them as a single system and considering their impact on each other. Typically, the various infrastructure components have interlinkages and interdependencies between each other. For example, the planning for road network expansion has to go hand in hand with planning the locations of junctions/roundabouts/flyovers, or planning the locations of drains along with planning of roads is important to avoid digging of roads later, which is quite a common feature in India. Similarly, location of bus or rail terminals requires good access, which is dependent on careful planning of other infrastructure. Therefore, it is essential to have integrated planning of transport infrastructure that takes account of and connects all these considerations and helps ensure development of the transportation network and land use in a coordinated manner. In this way, it ensures the most efficient use of public funds and avoids creating unintended impacts.

While integration is essential for efficient transportation networks, addressing first-mile and last-mile connectivity issues has become increasingly important. Many commuters face challenges reaching or leaving transit hubs like bus stops or metro stations due to a lack of pedestrian-friendly infrastructure, limited feeder services, and inadequate cycling or micro-mobility options. For instance, poorly maintained sidewalks, a lack of shared mobility services like e-rickshaws or bike-sharing, and minimal focus on accessibility for differently abled users hinder smooth transitions between transport modes. Coordinating these feeder systems with core transport schedules remains a challenge, often leading to long waiting times and a fragmented user experience.

Emerging technologies and innovative policies provide opportunities to address these gaps. Cities like Bengaluru and Delhi have started piloting first-mile and last-mile connectivity solutions, such as deploying electric feeder buses, integrating ridesharing services, and promoting cycling infrastructure. Additionally, digital platforms now allow users to plan trips seamlessly across modes, reducing reliance on private vehicles for short trips and improving public transport adoption.

To achieve integration, we need to consider a range of factors for both current and future demand, including:

1. Interaction of the transport network and land use
2. performance of the transport system
3. affordability
4. value for money
5. urban design
6. demand management
7. behavioral change
8. accessibility
9. safety
10. network management
11. balance of small-scale and large-scale improvements

12. coordination between different modes of transport, for example, cars, buses, rail, walking, cycles
13. key stakeholders involved in planning, operating, and using transportation networks
14. sequencing of works.

The benefits of an integrated approach to transport infrastructure planning are as follows:

1. improved transport efficiency and lower costs to deliver on economic growth and productivity goals.
2. better management of environmental and social impacts to enhance quality of life.
3. enhancements to the public realm to generate employment and community well-being.
4. efficient use of public funds
5. transport strategies and packages of activities are developed alongside land-use strategies and implementation plans.

While there are larger benefits of integration, we also need to be aware of the limitations associated with it:

1. the planning exercise becomes more complex and requires use of advanced tools and technologies to do it.
2. it requires substantial efforts in bringing various agencies together and develop consensus to do something, which is always very challenging at least in India.

The following are the areas of transit integration so as to achieve successful integration of transport infrastructures.

1. *Institutional Integration:* This refers to the creation of an organizational framework within which joint planning and operation of transit services can be carried out. Four types of organizational arrangements for implementing institutional integration can be made:
   a. *Tariff Association:* This is suitable only in situations where the partners do not compete and share no territory but rather make end-to-end connections. For instance, the suburban rail system in Mumbai is an example of intraurban transport. A passenger can purchase a single ticket covering western and central railways and pays no penalty for moving between the two railways.
   b. *Transit Communities:* These not only bind themselves to common traffic but coordinate routes and schedules and, if appropriate, pool or exchange some rolling stock. The railroads in the United States have long operated under such an arrangement.

c. *Transit Federations:* These establish a formal federated agency and delegate to it certain powers related to planning, tariffs, revenue distribution, and so on. The Munich (Germany) Transit Federation is an example.

d. *Mergers:* These are portions of companies that are merged into one firm, within which the companies either operate as subsidiaries or lose their identity altogether. Examples are transit services in London and Paris.

Of the several organizational alternatives, mergers appear to be the most effective in achieving a high degree of transit integration. A single authority is able to function more effectively and with greater flexibility than are associations of essentially independent public and private transit operators. Institutional arrangement in India for urban transport infrastructure has been extremely poor and still it is divided among multiple organizations in most of the Indian cities without any umbrella organization to control the quality and quantity of services. A few states have accepted and adopted to create the urban transport authorities with insignificant levels of success. The present level of institutional arrangements provides for competition across modes and infrastructure, whereas it should generally be among the providers of services under same mode. Individual institutions of different modes are stronger in existing set-ups than the umbrella organization, and it does not meet the objectives. In addition, in many cases, paratransit and other unconventional service supplies and operations are not safe and in violation of every rule and which are not seen and controlled by any institution. A massive overhaul of the institutional and management set-up with significant innovations in the institutions with defined tasks and targets is required. Qualified land use and transport planners are required to be at the helm of affairs to provide direction in the institutions, which is grossly lacking at this time. Multimodal integration is the key to efficiency in urban transport.

Complementariness of modes should be adopted as the first principle of planning multimodal transport for urban India. Seamless travel with the highest efficiency will be possible only when all the modes, including buses, metro, and paratransit, share the same objective and urban transport vision. Of course, they would have to be totally transparent in their business model so as to collectively adopt a fair division of responsibilities and revenues. In such a system, all effort of marketing, PIS, scheduling, etc. must be a coordinated effort so as to maximize the total business for the benefit of all partners. For successful multimodal integration to be achieved, integrated infrastructure, terminal facilities, etc. will have to be developed jointly by all the partner organizations to maximize the integration and the business. Institutional integration must prioritize first-mile and last-mile connectivity as a critical component of its framework. For instance, unified authorities must ensure feeder buses, shared mobility options like e-rickshaws, and bicycle-sharing systems are synchronized with main transit schedules and investments in real-time tracking and centralized route management systems can help reduce commuter waiting times and improve connectivity.

Case Example—Bengaluru**:** In India, the regulation of land transport is managed by multiple government agencies at the central, state, and local levels. The central government oversees national highways and railways, including metro systems, while state governments are responsible for state highways and local roads. In Karnataka, the Transport Department regulates MV usage and manages state-run transport corporations, while the Public Works Department oversees major district roads and highways. In Bengaluru, various agencies handle transport planning, infrastructure, and service provision. For instance, the Bengaluru Metro Rail Corporation Limited (BMRCL) operates the metro, and the BMTC manages bus services. Additionally, urban local bodies like the Bruhat Bengaluru Mahanagara Palike (BBMP) are responsible for road maintenance, while the Bengaluru Development Authority (BDA) prepares the city's Master Plan.

Recognizing the lack of coordination between these agencies, the MoHUA (2006) recommended the establishment of UMTAs in all million-plus cities to integrate urban transport planning and implementation. Since then, UMTAs have been formed in several cities, including Bengaluru, Chennai, Pune, Mumbai, Hyderabad, and Kochi. In 2007, Karnataka set up the Bengaluru Metropolitan Land Transport Authority (BMLTA) as Bengaluru's UMTA. The Directorate of Urban Land Transport (DULT) was also created to assess transport demand, recommend infrastructure projects, and assist local bodies in transport planning. However, BMLTA's effectiveness has been limited as many of its responsibilities have been taken over by other agencies. For example, the Comprehensive Mobility Plan for Bengaluru was prepared by DULT and BMRCL instead of BMLTA. Recognizing this gap, a Parliamentary Standing Committee suggested that UMTAs be granted statutory authority to strengthen their role.

To address these governance challenges, the Karnataka government introduced the Bengaluru Metropolitan Land Transport Authority Bill, 2022, which aims to provide BMLTA with legal backing and greater control over transport policy in the Bengaluru Urban Mobility Region. Under this bill, BMLTA will oversee transport planning, financing, regulation, and coordination across agencies. Its responsibilities include formulating policies for NMT, integrating various transport modes, engaging private sector players, and assisting the state government in mobility planning. A dedicated fund will be created to support these initiatives.

The BMLTA will consist of 33 members, including the Chief Minister as the chairperson and the State Transport Minister. Other members will include secretaries from key government departments, the Mayor and Chief Commissioner of BBMP, heads of planning and transport agencies like BDA, BMTC, and BMRCL, as well as representatives from law, academia, and civil society. An Executive Committee, chaired by the Chief Secretary, will assist BMLTA in implementing policies, while the Commissioner of DULT will serve as its CEO. The BMLTA is also tasked with preparing a Comprehensive Mobility Plan, revised every five years, to guide infrastructure development, including public transport routes, road networks, pedestrian pathways, and traffic management strategies. Additionally, a City Mobility Investment Program will be formulated every two to five years to finance urban transport projects, while Annual Implementation Plans will be created by relevant agencies to execute transport initiatives.

A key provision in the bill mandates that urban planning authorities must align their Master Plans with BMLTA's transport strategies. While this integration is essential for seamless urban mobility, concerns have been raised that BMLTA's role in land-use planning could give it excessive influence over urban development decisions. Traditionally, Master Plans are designed considering economic, social, and environmental factors, with transport being one of many components. Granting BMLTA overriding authority in this process could lead to transport-centric planning that may overlook broader urban development priorities.

A major concern is the limited representation of local bodies within BMLTA. The Bengaluru Metropolitan Area includes multiple urban and rural jurisdictions beyond BBMP's limits, yet these areas lack direct representation in BMLTA's decision-making process.

Case Example—Chennai: The Chennai Unified Metropolitan Transport Authority (CUMTA) serves as the nodal agency for urban transport in Chennai. It was established in November 2010 by the Government of Tamil Nadu through a statutory act. Its primary role is to enhance transport infrastructure and ensure safe, efficient, and sustainable mobility across the Chennai Metropolitan Area (CMA). As an institutional coordination mechanism, CUMTA integrates transport planning and service delivery across multiple agencies, including bus, train, and metro services, to provide seamless urban mobility.

CUMTA's jurisdiction covers the entire Chennai Metropolitan Area, which has expanded from 1189 to 5094 sq. km to accommodate the city's growing transport needs. The rules governing CUMTA were officially notified in January 2019, and its statutory framework was strengthened through an amendment in January 2021. The authority is chaired by the Hon'ble Chief Minister of Tamil Nadu, with the Minister in charge of Housing and Urban Development serving as the Vice-Chairperson. CUMTA works with multiple agencies to implement integrated urban transport solutions, including street development, multimodal transport integration, and common mobility payment systems.

The vision of CUMTA is to facilitate the seamless movement of people through an integrated, sustainable, and resilient transport ecosystem. Its mission focuses on ensuring safe and efficient mobility while promoting economic growth and environmental sustainability. One of CUMTA's key responsibilities is multimodal integration, ensuring that various modes of transport—bus, metro, and suburban rail—are physically, financially, and operationally integrated. It oversees the approval of plans, regulates multimodal initiatives, monitors fund utilization, and facilitates interagency coordination. A dedicated multimodal integration sub-committee is tasked with identifying integration projects, ensuring seamless fare and service coordination, and overseeing the implementation of critical initiatives such as Mobility as a Service (MaaS) and common digital payment systems.

CUMTA also plays a crucial role in advancing digital urban mobility solutions under the "Digital Chennai" initiative. This includes setting up an integrated mobility data platform, fostering interagency coordination for digital infrastructure, and developing digital solutions such as a centralized urban mobility management system. The Digital Chennai sub-committee oversees the development of an Integrated Urban Development Platform (IUDP) and an Urban Project Planning and Management (UPPM) tool, aimed at optimizing urban service delivery and improving coordination among various agencies.

Urban mobility resilience is another key focus area for CUMTA, ensuring that Chennai's transport infrastructure can withstand disasters and emergencies. The authority is responsible for conducting risk assessments, formulating contingency plans, and ensuring interagency coordination for emergency management, recovery planning, and resilience-building. The Urban Mobility Resilience sub-committee monitors compliance with resilience standards and integrates disaster preparedness into transport infrastructure planning.

In addition to multimodal integration and resilience planning, CUMTA is actively involved in promoting road safety and NMT infrastructure. The Road Safety and NMT sub-committee is responsible for reviewing and enforcing road safety policies, improving pedestrian and cycling infrastructure, and aligning transport safety initiatives with national and global standards. It also monitors the implementation of measures aimed at reducing traffic fatalities and ensuring safer roads for all users.

Despite its potential, CUMTA faces challenges such as limited autonomy, bureaucratic delays, and overlapping responsibilities with other agencies. Strengthening its authority, ensuring financial independence, and improving coordination with local bodies will be crucial for achieving its vision of an integrated, efficient, and sustainable urban transport system in Chennai.

2. *Physical Integration:* This refers to the provision of jointly used facilities and equipment. It is a critical component in ensuring effective first-mile and last-mile connectivity, which directly impacts the overall accessibility and usability of public transport systems. By incorporating facilities that cater to both main transit services and feeder systems, physical integration can bridge the gaps between modes and improve commuter experience. There are various techniques of physical transit integration; some of them are as follows:
   a. *Intermodal Terminals:* Transfer between modes of transit service is facilitated by intermodal terminals, often described as transportation centers. The most highly developed of these facilities accommodate commuter rail lines, RRT lines, light rail and streetcar lines, and bus services, with facilities of transfer between themselves or from dial-a-ride or circulation feeder services, taxis, or private vehicles. To address

first- and last-mile connectivity, these terminals must include well-designed feeder systems like shuttle buses or e-rickshaws to link nearby neighborhoods to main transit hubs. Parking accommodation is provided to encourage park-and-ride travel. Last-mile connectivity can be further enhanced by providing pick-up and drop-off zones (kiss-and-ride), along with secure spaces for bicycles and pedestrians. Integrating these services into a unified system helps streamline travel and reduces dependency on private vehicles. Where single-fare systems for all transit modes have been established, passengers move freely among different services without being stopped by barriers or turnstiles.

   b. *Transit Shelters:* Transit shelters range from simple weather-protection structures on surface transit routes to mini-terminals at important stops and transfer points. In the context of first-mile and last-mile connectivity, these shelters should be designed to include facilities that connect passengers to nearby residential or commercial areas. The more complex facilities may provide automatic ticket vending machines, free direct-line telephone connections to a centralized information service, locater maps, posted routes and schedules, and promotional material on reduced-fare multiride passes or special excursions.

3. *Network Integration:* This involves the application of management techniques to optimize the allocation of transit resources and coordinate services. The techniques of operational integration include (among others) the following:
   a. *Coordinated Routing and Scheduling:* Routing and scheduling for high-capacity, long-haul modes, such as commuter rail and RRT, are considered as the main system and buses act as a feeder to the rail system. Accordingly, the integrated route network is planned by generating feeder bus routes for each rail station. Also, the schedules of trains and buses are coordinated to minimize transfer time between the two modes.
   b. *Rationalization of Redundant Services:* The wasteful duplication of transit service by competing systems is eliminated, and resources are redeployed to reduce headways on existing routes and extend services into new areas.
   c. *Information Integration:* Information on routes, schedules, fares, and transfer points for all transit modes and services throughout the urban area is provided by a centralized source. A centralized system that provides real-time updates on routes, schedules, and fares across all modes of transport allows commuters to plan their trips seamlessly from start to finish. Information services include route maps, timetables, fare schedules, promotion materials, uniform street signs, vehicle identification, displays at stops, transfer points, and major stations, as well as telephone inquiry answering service. Providing integrated information during every leg of the journey is important to make them attractive (Figure 3.5).

d. *Fare Integration:* A single, area-wide fare structure is established that permits riders to purchase one ticket at the beginning of the trip and transfer freely among all modes or lines of service within the system. For example, in Berlin, Germany (Figure 3.6), one can buy a transit ticket and travel seamlessly between metro (surface as well as underground), bus, tram, and ferry.

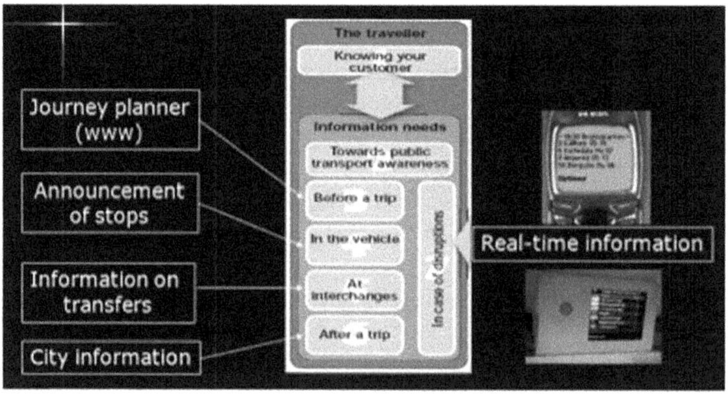

FIGURE 3.5 Information Integration

FIGURE 3.6 Fare Integration in Berlin, Germany

3.5 FUNDING PATTERN

The development of public transport has been widely recognized to achieve environmental, social, and economic sustainability in cities in developed and developing countries (Cervero, 1998; Vuchic, 2005). However, the emphasis on environmental and social sustainability of public transport is frequently compromised due to a focus on financial sustainability (Buehler and Pucher, 2011). Comprehensive public transport infrastructure is expensive to build and costly to operate (White, 2002; Ubbels and Nijkamp, 2002). Moreover, the costs of public transport agencies have increased significantly with the broadening of policy goals, and, therefore, many public transport systems are struggling for funding.

Traditional methods of funding, such as public sector funding, subsidies, and revenue from fares, are not sufficient to make improvements to public transport infrastructure and innovative funding mechanisms are, therefore, demanded. This problem is made more challenging because public transport is generally not a profitable investment for the private sector. Often public transport systems use low fares as a means of promoting their use which require government subsidies. There are different methods of funding options, and they are mentioned below:

a. Traditional public transport funding options
b. Innovative options to fund public transport.

3.5.1 Traditional Public Transport Funding Options

Traditional options used to fund public transport include subsidies, government funding, and fare box revenue.

3.5.1.1 Subsidies

Subsidies are very simple to understand. Basically, a government allocates funds paid in income or corporate taxes to public transport systems that usually run at a financial loss the subsidies ensure that the systems can continue to operate even if they cannot be profitable (Ubbels and Nijkamp, 2002; Ubbels et al., 2001). Subsidies have been used to increase public transport use in order to reduce environmental externalities like air and noise pollution, GHG emissions, and congestion. In this situation, benefits are broadening to those who do not even use the public transport services (Serebrisky, 2009). Although subsidies create social and environmental benefits, the economic costs are progressively increasing. For example, the provision of low fares through subsidies on public transport in Madrid has decreased the amount of revenue from users yet operational costs are increasing (Manuel et al., 2009). Subsidies can also lead to inefficiencies and are viewed as an inferior good.

In summary, subsidies have an important role in public transport, but it is not sustainable to expect long-term funding from subsidization to increase or even remain at the same level. This is due to increasing competition for government funding from the priority areas of education and health.

3.5.1.2 Fares

The collection of fares allows public transport operators to reclaim the cost of operating the system. However, the fare box revenue is usually insufficient to pay for both the capital cost and running expenses of a modern mass transit system (Wetzel, 2006). For example, in the United States, fares pay for an average of only 35–42% of operating costs (but not capital costs) of public transport, while in Europe fares cover an average 45–48% of total operating costs (Pucher et al., 1983; Reeven, 2008). Despite these trends, efficient public transport systems perform very well. The Hong Kong and Singapore public transport systems operate at profit while German cities cover two-thirds of their operating costs from fare box revenue.

In order to plan fares for a given transit system, it is necessary to define what the fares should achieve. The basic objectives for a fare system may be the following:

1. to attract the maximum number of passengers
2. to generate the maximum revenue for the transit agency
3. to achieve specific goals, such as increasing the mobility of the labor force, students, or seniors, improving access among certain areas, and promoting the use of a more efficient transit mode.

The common requirements and constraints that must often be considered in setting up fares include the following:

1. Elasticity of demand may limit the choice of fare level and structure; the quality of service and price of competing modes play a major role in passengers' willingness to use transit at a given fare level.
2. Equality among transit travelers in terms of value of services (trip length, local or express, comfort level, safety, and security) versus the fare paid by different user groups. This may be an important factor for individual users, as well as for groups of population.
3. Social and political aspects or the need for service and the ability to pay by different population groups often play an important role.
4. Fares should be understandable and convenient for passengers to pay.
5. Fare type should allow easy and low-cost collection and control by the agency.

The process of fare collection—the locations and methods of collecting fares and control of payments by passengers—has a major impact on passenger convenience, on vehicle standing (dwell) times at transit stops, and on the cost of collecting revenue. Street transit, mostly bus operations, is particularly affected by the delays fare collection may cause at transit stops. When fares are collected in stations, usually rail, fare collection has no impact on boarding, but the cost of collection may be high if it involves personal or complicated equipment. Self-service systems and the use of smart cards may avoid most of these disadvantages when they can be applied.

3.5.1.3 Fare Structure

Fare structures are classified on the basis of the relationship between fare amount and distance traveled on a transit line. Using this criterion, there are two main structures:

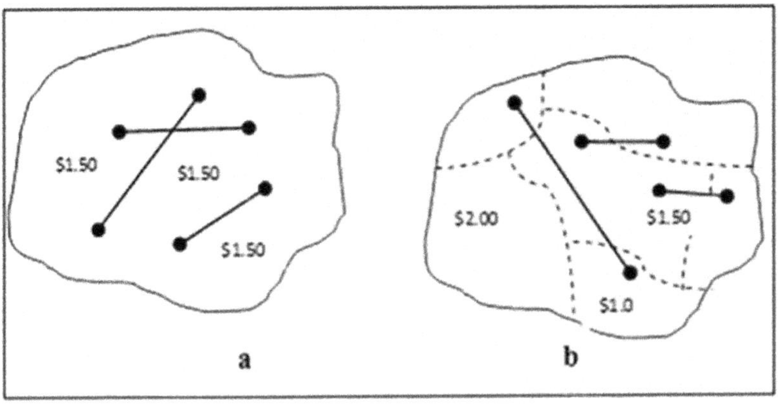

FIGURE 3.7 Schematic Presenting of Alternative Fare Structures: (a) Flat Fare; (b) Zonal Fare

flat fare and graduated fare. The latter can be further subdivided into zonal and sectional fare. Flat fare is a constant amount independent of trip length. In contrast, zonal and sectional fares change with distance traveled on the transit network. The three types of fare structures are shown schematically in Figure 3.7.

3.5.1.4 Flat Fare

As the name implies, flat fare is constant regardless of the distance a passenger travels on a single vehicle or throughout the transit network. The flat fare is the simplest possible fare; therefore, it is easy for passengers to understand and use. If it is used for individual lines, it is usually collected at fare gates in a station or upon boarding the transit vehicle. Thus, supervision of payment is easy, the work of fare collectors or vehicle operators is simple, and the boarding process is rather fast.

The basic disadvantage of a flat fare is that it is a fixed amount that does not in any way reflect the quantity of service that the rider receives: a passenger traveling three street blocks pays the same amount as a passenger traveling 8 or 10 km.

3.5.1.5 Graduated Fares

To avoid the inherent limitation of the flat fare, the inequity of short-distance riders actually subsidizing long-distance riders and the inability to collect additional revenue from the latter group, graduated fares can be used. Graduated fares can include many variations, but as noted above, there are two basic types: zonal fare and sectional fare. They are described in the following two sections.

3.5.1.6 Zonal Fare

The simplest method for charging graduated fares proportionally to the lengths of passenger trips is achieved by dividing the city or urban area into fare zones:

sometimes two (central and outlying), sometimes more. Then one fare applies to passenger trips wholly within a single fair zone, a higher fare applies to passenger crossing from one zone to another, an even higher applies to passengers whose trip is across two fare zone boundaries, etc.

The main advantage of a zonal fare structure is that it provides a uniform base fare for a given small geographic area, as well as a correspondingly higher amount of revenue from longer trips. In comparison with a flat fare, a zonal fare has the advantage of charging increasing amounts for longer trip lengths. This makes the fare more equitable and acceptable to the users. Consequently, zonal fares usually result in both attraction of higher patronage and collection of greater revenue than flat fares.

3.5.1.7 Sectional Fare

Another way of making fares related to travel distance is obtained by dividing transit lines into sections. The fares then increase with the number of sections traveled. As sections are usually shorter and fare increments smaller than zone diameters, sectional fares are even more closely related to travel distances than zonal fares. On the other hand, sectional fares are correspondingly more complicated to compute, collect, and control than zonal fares. Consequently, sectional fares often result in longer collection times, require more personnel, and allow greater opportunity for evasion of full fare payment. The fare system of most of bus operators in India like BMTC is based on the sectional fare system.

To recapitulate, both zonal and sectional fares have amounts that vary with trip lengths, but they are computed differently: zonal fares are determined on the basis

TABLE 3.7
Comparison of Trade-offs among Alternative Fare Structure Types and Their Optimal Applications

Characteristics		Types of Fares		
		Flat	**Zonal**	**Sectional**
Important characteristics	Equity	Poor	Good	Very good
	Passenger attraction	Good	Very good	Very good
	Revenue collection	Variable	Good	Very good
	Simplicity of collection	Excellent	Fair–good	Poor
	Simplicity of control	Excellent	Fair	Poor
	Simplicity for passengers	Excellent	Fair–good	Poor
Desired conditions	Line length	Short (<5 km)	Medium	Long
	Network type	Ubiquitous	Divisible in zones	Long line
	Travel distance	Short	Variable	Variable

of geographically defined zones, while sectional fares are based on the distance traveled in a single transit line.

3.5.1.7.1 Comparison of the Three Basic Fare Structures

The preceding description of the three basic types of fare structures shows that each has its advantages and shortcomings. In general, there are major differences between flat and graduated fares and minor differences between the two types of graduated fares, zonal and sectional. The main differences are summarized in Table 3.7.

3.5.1.8 General Taxes

General taxation is the most widely used traditional form of funding for public transport (Mills, 1991; Ubbels and Nijkamp, 2002). Although subsidies are typically funded from general taxation, it is important to recognize that general taxes are also used for wider infrastructure development (Mills, 1991; Ubbels and Nijkamp, 2002). The most widely used general taxes include income tax rates (property taxes) and goods and services taxes.

On the other hand, general tax has been perceived to be an appropriate approach to mitigate unsustainable transport behavior as this form of funding message is effective in discouraging private car use. This is because, using taxes to fund public transport means that the majority of people, regardless of whether or not they use public transport, are paying for and making the service cheaper for patrons (Buehler and Pucher, 2011).

3.5.2 Innovative Options to Fund Public Transport

Increasingly, fare box revenue, subsidies, and government funding are not enough to build public transport infrastructure and operate the system. Therefore, local-based innovative charges and taxes have been introduced where some or all of the revenue is directed to public transport (Enoch et al., 2005; Tsukada and Kuranami, 1994). These innovative funding options for public transport are assessed under four headings: beneficiary pays development taxes/funds; polluter pays carbon taxes/funds; public transport operation taxes/funds; and private sector and other transport taxes and/or funds. Under each category a variety of options are discussed.

3.5.2.1 Beneficiary Pays Development Taxes/Funds

Beneficiaries paying public good taxes have traditionally been used to fund services such as police, fire services, and ambulances which cannot be provided on a market-exchange basis. Many projects and some elements of public transport are also funded on this basis (Enoch et al., 2005). Within this beneficiary pay category a number of options are available, many with their own subcategories. These are as follows:

1. Property taxes/rates
2. Land acquisition along public transport routes
3. Smart tax
4. Off-set charges
5. Student surcharges

3.5.2.2 Polluter Pays Carbon Taxes/Funds

1. Fuel charges
2. Regional fuel tax
3. Road pricing
4. Parking charges and fines
5. Fee on parking buildings
6. Vehicle registration fee
7. Vehicle kilometers traveled (VKT) taxes
8. Emission charges

3.5.2.3 Public Transport Operational and Infrastructural Taxes/Fund

There are a number of options available to increase funds from public transport operations and infrastructure. These include the following:

1. Reprioritizing transport funding
2. Public transport trust fund
3. Railway development fund
4. Improving public transport network
5. Multimodal passes
6. Higher usage through modern services
7. Leasing public transport workshops
8. Advertising
9. Naming of routes and bus stops (naming rights)
10. Tourism
11. Fee on HOV lane use

3.5.2.4 Private Sector Taxes/Funds

A number of options are available to increase funds from private sector organizations. These are as follows:

1. Public–private partnership (PPP)
2. Privatization
3. Competitive tendering

3.5.2.5 Other Taxes/Funds

1. Visitor tax
2. Payroll tax
3. Cross-utility financing
4. Consumption tax
5. Luxury car tax

SUMMARY

This chapter mainly covers the detailed concept of the public transport scenario in a developing country like India. The current perception of India is well explained by taking Bengaluru city as an example and implementing the SERVQUAL framework for compilation and analysis. In addition, the current planning and organization of Bengaluru city is covered by taking the performance of BMTC from last several years. The chapter also deals with the LOSs and coordination issues by taking Bengaluru and Chennai as case studies. It also sheds some light on the different funding patterns adopted and how they are performing.

4 Public Mass Transport Planning in Urban Areas

4.1 URBAN TRAVEL DEMAND ESTIMATION FOR DIFFERENT CATEGORIES OF URBAN AREAS FOR THE NEXT TWO DECADES

The development of the 3Cs (continuing, cooperative, and comprehensive) is very important in the planning and development of sustainable transportation infrastructure.

1. Continuing specifies that the planning process be ongoing, frequently reevaluating and updating the transit plan to reflect changes in the urban area.
2. Cooperative is defined as the need for coordination not only between the various levels of government (local, state, and central) but also among individual agencies at the same level. It also implies cooperation between disciplines during the planning process and mandates that planning teams be interdisciplinary.
3. Comprehensive implies the set of the following ten elements for which inventories and analyses are required:
    a. Economic factors affecting development
    b. Population
    c. Land use
    d. Travel patterns
    e. Existing transportation facilities
    f. Terminal and transfer facilities
    g. Traffic control features
    h. Zoning ordinances, subdivision regulations, building codes, etc.
    i. Financial resources
    j. Numerous qualitative elements, including social and community value factors, such as preservation of open space, environmental amenities, and aesthetics.

For any metropolitan planning organization (MPO) like the development authorities in Indian cities will typically include the sustainable transport infrastructure planning as shown in Figure 4.1. As can be seen in the figure, any such organization should have a planning works program, based on both short-term and long-term planning elements. The short-term planning elements, generally known as transportation system management (TSM) elements, are basically meant for efficient use of

DOI: 10.1201/ 9781003740247-4

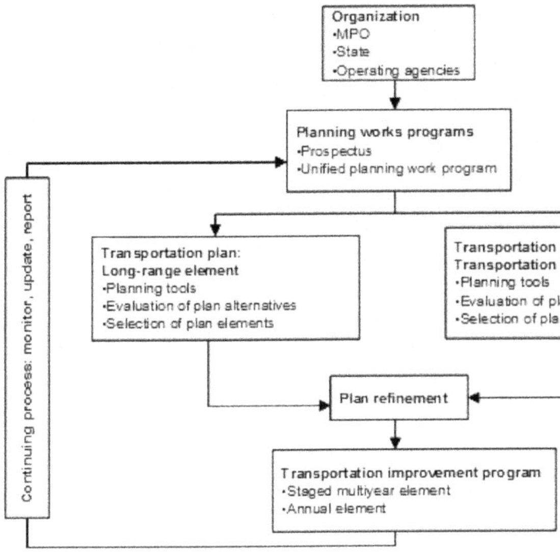

FIGURE 4.1 Planning Process of an Organization

existing and proposed infrastructure. Together with short- and long-term elements, the organization will have annual and multiyear improvement programs, which will be frequently updated and monitored from time to time.

Figure 4.2 gives the basic steps in a comprehensive urban transport infrastructure planning process. The same are described as follows:

1. Definition of goals and objectives for the transportation system in the future.
2. Collection of inventories: data about the existing city and its transportation system.
3. Forecasts of changes and conditions in the selected target year for plans.
4. A set of criteria for plan evaluation derived from the goals and objectives.
5. Development of alternative infrastructure plans for the projected future conditions, meeting the defined goals.
6. Technical elaboration and testing of alternative plans, considering their impact on projected demand and urban development.
7. Comparative evaluation of alternative infrastructure plans using the set of goal-based criteria and public hearings, resulting in selection of the preferred plan.
8. Finalization of the selected plan and preparation for its implementation.

This is a conceptual organization of a planning process that includes the entire planning from data collection to the production of the final plan and its implementation. Each of these steps further consists of data handling, modelling, technical analysis, etc. Figure 4.3 shows the planning process with more technical details.

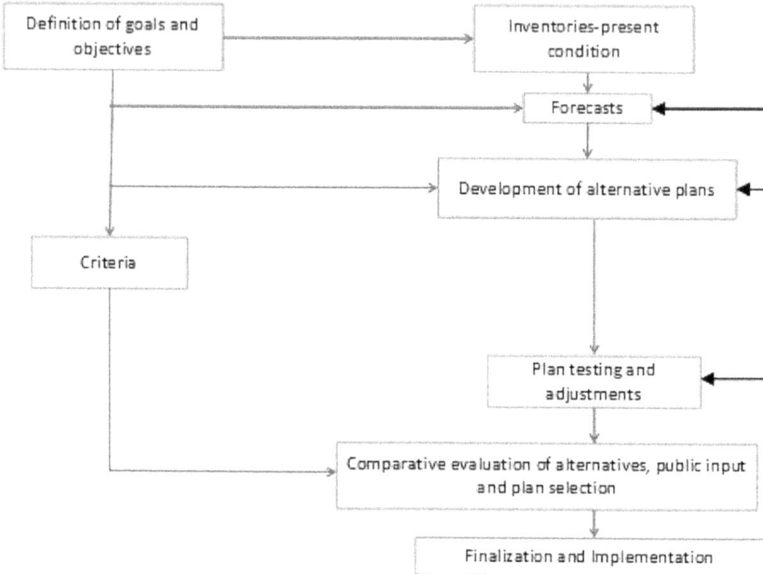

FIGURE 4.2 Sequence of Basic Steps in Comprehensive Urban Transportation Infrastructure Planning

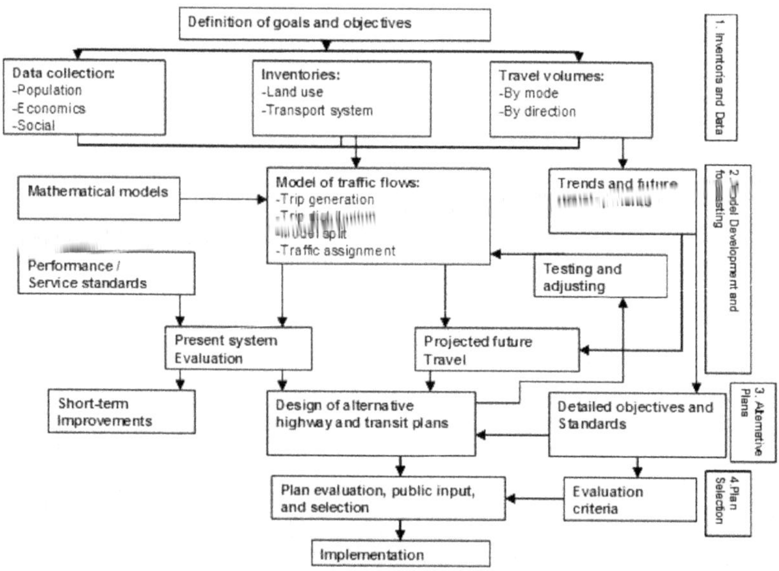

FIGURE 4.3 Technical Procedure of Long-Range Urban Transportation Infrastructure

The data collection usually consists of collecting comprehensive information about the present transportation system and urban area. The following are the main categories of the essential data required for sustainable infrastructure purpose:

1. *Land-Use Characteristics*
   a. Population of localities
   b. Residential areas
   c. Commercial areas
   d. Educational areas
   e. Recreational areas
   f. Developable areas
   g. Other land uses

   For example, Figure 4.4 shows the proposed land-use map of Bengaluru city for 2031
2. *Network Characteristics*
   a. Roads by classification
   b. Road properties
   c. Road geometry
   d. Railway system properties
   e. Other network properties

   Figure 4.5 shows the network map of the existing and proposed Bengaluru Metropolitan Region for the year 2023
3. *Topographical Features*
   a. Contours
   b. Soil classification
   c. Water bodies
   d. Drainage patterns
   e. Quarry identification
   f. Forest cover
   g. Identification of obligatory points
4. *Socioeconomic Data*
   a. Demographic information
   b. Income levels
   c. Vehicle ownership levels
   d. Occupation
   e. Education
   f. Family structure
   g. Employment
5. *Activity Information*
   a. Industrial employment
   b. Educational employment
   c. Administrative employment
   d. Wholesale and retail employment

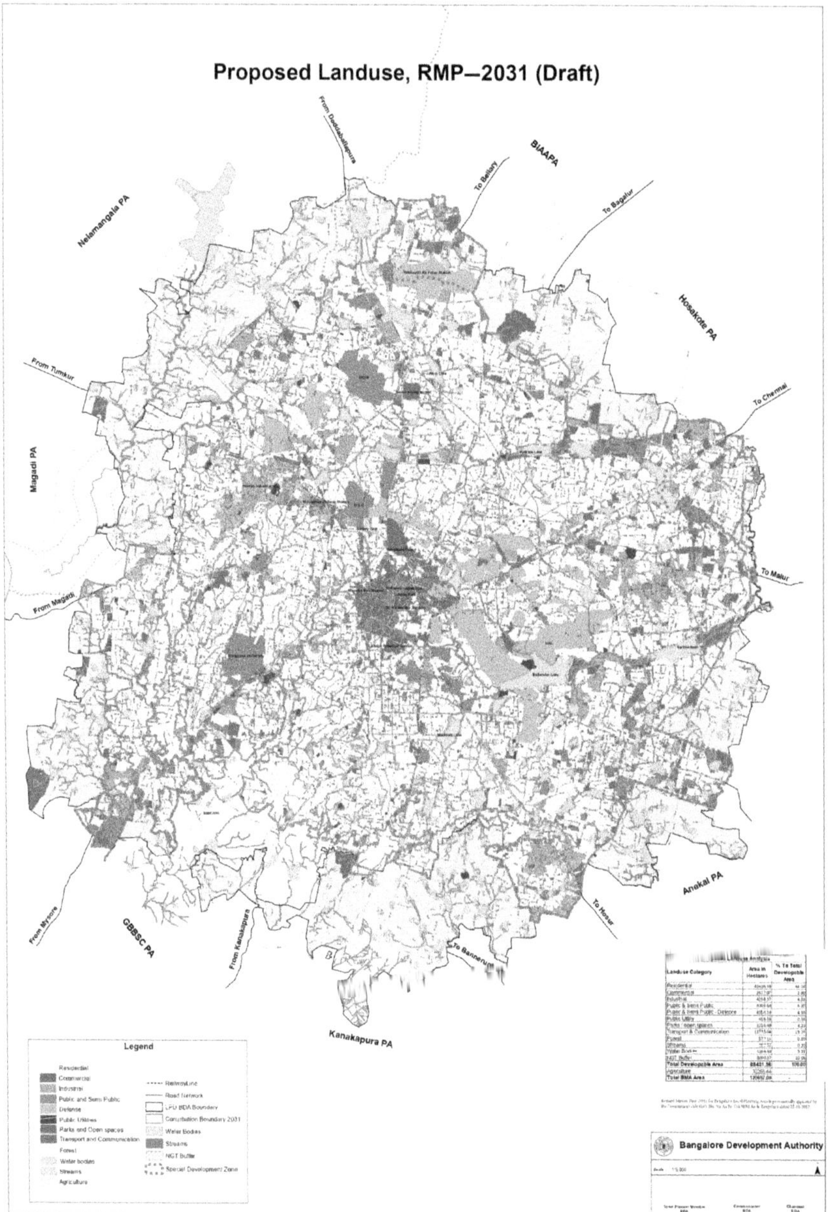

FIGURE 4.4 Proposed Land-Use Map of Bengaluru City for 2031

Source: Bengaluru Development Authority

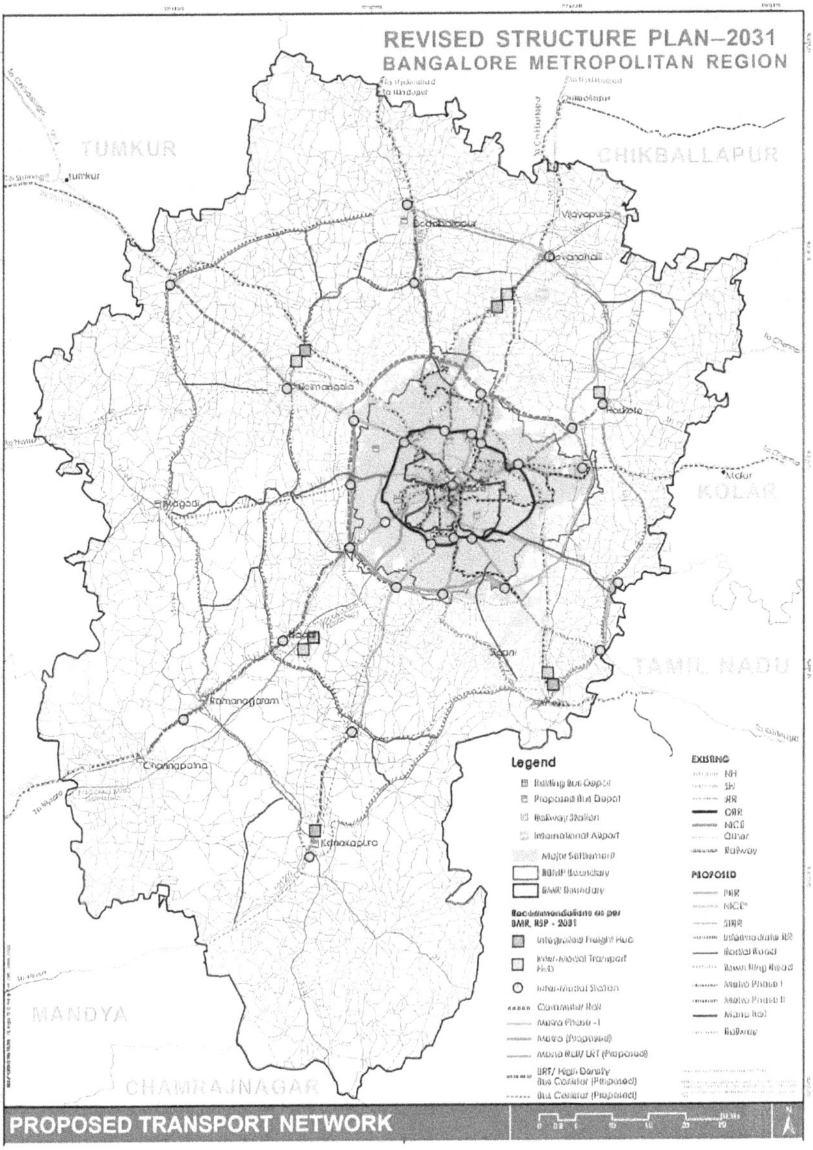

FIGURE 4.5 Network Map of Bengaluru Metropolitan Region

Source: Bengaluru Development Authority

   e. Service employment
   f. Informal sector employment
6. *Transport System Characteristics*
   a. Capacity
   b. Speeds
   c. Frequency
   d. Comfort
   e. Reliability
   f. Time of travel
   g. Modal integration

The purpose of the travel demand modeling phase of the urban transportation infrastructure planning process is to perform a conditional prediction of travel demand in order to estimate the likely transportation consequences of several transportation alternatives that are being considered for implementation. This prediction is also conditional on a predicted target-year land-use pattern. The major components of travel behavior were identified as follows:

1. The decision to travel for a given purpose (trip generation).
2. The choice of destination (trip distribution).
3. The choice of travel mode (mode choice).
4. The choice of route or path (route assignment).

The four-step modeling process was originally developed in connection with the planning of major highway facilities during the 1950s and 1960s. At first glance, the process may appear to have remained unchanged since that time. In reality, however, it has undergone significant modification in response to an improved understanding of travel behavior by modelers, the need to address emerging policy questions, and advances in computational technology. More powerful personal computers allowed the specification of more complex and detailed models. This evolutionary step will undoubtedly continue.

4.1.1 Structure of the Classical Transport Model

Years of experimentation and development have resulted in a general structure which has been called the classic transport model. This structure is, in effect, a result of practice in the 1960s but has remained more or less unaltered despite major improvements in modeling techniques during the 1970s and 1980s.

The classic model is presented as a sequence of four sub-models. These are as follows:

1. *Trip Generation:* It determines the frequency of origins or destinations of trips in each zone by trip purpose, as a function of land uses and household demographics, and other socio-economic factors.

2. *Trip Distribution:* It matches origins with destinations and is concerned with the estimation of target year trip volumes that interchange between all pairs of zones i and j, where i is the trip-producing zone and j is the trip-attracting zone of the pair.
3. *Mode Choice:* This step computes the proportion of trips between each origin and destination that use a particular transportation mode.
4. *Route Assignment:* In this step, trips between an origin and destination are allocated by a particular mode to a route.

4.1.2 Trip Generation

Trip generation is the process by which measures of urban activity are converted into number of trips. For example, the number of trips that are generated by a shopping center is quite different from the number of trips generated by an industrial complex that takes up about the same amount of space. In the trip generation, the planner attempts to quantify the relationship between urban activity and travel.

The study area is divided into zones for analysis purposes. After trip-generation analysis, the planner knows how many trips are produced by each zone and how many are attracted by each zone. In addition, the planner knows the purposes for the trips—the trips are put into several categories, like trips from home to work, or home to shop, or home to school. This categorization is necessary because each trip purpose reflects the behavior of the trip maker. For example, school trips and work trips are pretty regular; shopping and recreation trips are less so. The three most common mathematical formulations of trip generation are multiple linear regression technique, trip rate-analysis technique, and cross-classification technique. Before going into these models, let us have a look at some of the basic definitions.

1. *Journey:* This is a one-way movement from a point of origin to a point of destination.
2. *Home-Based (HB) Trip:* This is one where the home of the trip maker is either the origin or the destination of the journey.
3. *Non-Home-Based (NHB) Trip:* This, conversely, is one where neither end of the trip is the home of the traveler.
4. *Trip Production:* This is defined as the home end of an HB trip or as the origin of an NHB trip.
5. *Trip Attraction:* This is defined as the nonhome end of an HB trip or the destination of an NHB trip.
6. *Trip Generation:* This is often defined as the total number of trips generated by households in a zone, be they HB or NHB.

4.1.2.1 Multiple Linear Regression Technique

Transportation studies have shown that residential land use is an important trip generator. Also, non-residential land use in many cases is a good attractor of trips. A typical equation connecting estimated trips generated by a residential zone could be:

$$Y = A + B_1X_1 + B_2X_2 + B_3X_3$$

where

Y = trips/household
X_1 = car ownership
X_2 = family income
X_3 = family size

Model parameters and variables vary from one study area to another and are established by using base-year information. Once the equations are calibrated, they are used to estimate future travel for a target year. In developing regression equations, the following is assumed:

1. All the independent variables are independent of each other.
2. All the independent variables are normally distributed.
3. The independent variables are continuous.

It is not possible for the planner to conform to these specifications, and because of this departure, regression analysis has come under considerable criticism. A typical regression equation may appear as follows:

$$Y = 0.0649X_1 - 0.0034X_2 + 0.0063X_3 + 0.9486X_4$$

To derive an estimate of a zone for a future year, appropriate estimates of X_1, X_2, X_3, and X_4 are substituted, and the equation is solved. The quality of fit of a regression line determined by coefficient of determination (goodness of fit) represented by R^2, the value being between 0 and 1. The closer R is to 1, the better the linear relationship between the variables.

The multiple regression technique is appealing to transportation analysts because it is easy to determine the degree of relationship between the dependent and independent variables. Also, it is possible to determine the ability of the equation to predict accurately. Apart from the coefficient of multiple correlation referred to before, it is the practice to determine the standard error of estimate (S_e), which is a measure of the deviation of, say, observed trips from values predicted by the equation. In addition, the partial correlation coefficients (r_j) of each of the independent variables is calculated. It provides the relationship between the dependent variable (y) and the particular independent variable (X_i) under investigation. The *t*-test may be used to determine whether an estimated regression coefficient is significant. In summary, if

$$Y_e = A + B_1X_1 + B_2X_2 + B_3X_3 \ldots\ldots\ldots B_n + X_n$$

then

$$R^2 = \frac{\Sigma(Y_e - \mu_y)^2}{\Sigma(Y_i * \mu_y)^2}$$

$$S_e = \frac{\Sigma(Y_e)^2}{N-(n+1)}$$

$$S_{Bn} = \frac{(S_e)^2}{S_{xi} * N(1-((R_{xi})^2)}$$

$$t = \frac{B_n}{S_n}$$

where

Y_e = estimated magnitude of the dependent variable
μ_Y = mean of the dependent variable
R^2 = coefficient of multiple determination
R = multiple correlation coefficient
S_e = standard error of estimate
S_{Bi} = standard deviation of the independent variable X_i
R_{x2} = coefficient of multiple correlation between X_i and all other independent variables

4.1.2.2 Trip-Rate Analysis Technique

Trip-rate analysis refers to several models that are based on the determination of the average trip-production or trip-attraction rates associated with the important trip generators within the region. Table 4.1, for example, displays the trip-generation rates associated with various land-use categories in downtown Pittsburgh, which were obtained by one of the very first major urban transportation studies.

This table includes production rates by residential land uses and attraction rates by several non-residential land uses. Care must be exercised to apply trip-rate models in the same context that they were calibrated. In this case, the rates represent person-trips (rather than vehicle trips) per thousand square feet of each land use.

TABLE 4.1
Floor-Space Trip-Generation Rates Grouped by Generalized Land-Use Categories

Land-Use Category	Thousands of Square Feet	Person-Trips	Trips per Thousand Square Feet
Residential	2744	6574	2.4
Commercial			
Retail	6732	54,833	8.1
Services	13,506	70,014	5.2
Wholesale	2599	3162	1.2
Manufacturing	1392	1335	1.0
Transportation	1394	5630	4.0
Public buildings	2977	11,746	3.9
Total average	31,344	153.294	4.9

4.1.2.3 Cross-Classification Technique

Cross-classification (or category analysis) models may be thought of as extensions of the simple trip-rate models. Although they can be calibrated as area- or zone-based models, in trip-generation studies they are almost exclusively used as disaggregate models. In the residential-generation context, household types are classified according to a set of categories that are highly correlated with trip making. Three to four explanatory variables, each broken into about three discrete levels, are usually sufficient. Typically, household size, automobile ownership, household income, and some measure of land development intensity are used to classify household types.

The trip rates associated with each type of household are estimated by statistical methods, and these rates are assumed to remain stable over time. Table 4.2 presents a cross-classification table that shows the calibrated nonwork HB trip-production rates for various types of households defined by (1) four levels of household size (i.e., number of persons per household) and (2) three levels of car ownership (i.e., dwelling units per acre), a surrogate for accessibility to nonwork activities (e.g., shopping, entertainment). Using the calibrated cross-classification table of Table 4.2, we estimate the total nonwork HB trips that the zone will produce during a typical target-year day.

Example 1: An urban zone contains 200 acres of residential land, 50 acres devoted to commercial uses, and 10 acres of park land. Table 4.3 presents the zone's expected household composition at some future (target) year. Using the calibrated cross-classification table of Table 4.2, the total nonwork HB trips that the zone will produce during a typical target-year day can be estimated. The rates are given as trips per household per day.

Solution: The total productions are estimated by summing the contribution of each household type:

$$p_1 = \sum_h N_h R_h$$

where N_h and R_h are the number of households of type h and their corresponding production rate. For example, the 300 single-person one-car households contribute

TABLE 4.2
Total HB-Nonwork Trip Rates

Area Type	Vehicles Available per Household	Persons per Household			
		1	2,3	4	5+
1. Urban: high density	0	0.57	2.07	4.57	6.95
	1	1.45	3.02	5.52	7.90
	2+	1.82	3.39	5.89	8.27
2. Suburban: medium density	0	0.97	2.54	5.04	7.42
	1	1.92	3.49	5.99	8.37
	2+	2.29	3.86	6.36	8.74
3. Rural: low density	0	0.54	1.94	4.44	6.82
	1	1.32	2.89	5.39	7.77
	2+	1.69	3.26	5.76	8.14

TABLE 4.3
Persons per Household

Vehicles per Household	Persons per Household			
	1	2, 3	4	5+
0	100	200	150	20
1	300	500	210	50
2+	150	100	60	0

Source: Papacostas and Prevedouras (2002)

(300) (1.45) = 435 nonwork HB trips per day. Summing over all household types gives $p_1 = 5760$ trips per day.

4.1.3 Trip Distribution

Trip generation models can be used to estimate the total number of trips emanating from a zone and those attracted to each zone. Generations and attractions provide an idea of the level of trip making in a study area, but this is often not enough for modeling and decision-making. What is needed is a better idea of the pattern of trip making, from where to where do trips take place, the modes of transport chosen, and the routes taken. Figure 4.6 shows a desire line diagram showing trips made between different places in MMR.

4.1.3.1 Growth-Factor Methods

Let us consider first a situation where we have a basic trip matrix t, perhaps obtained from a previous study or estimated from recent survey data. We would like to estimate the matrix corresponding to the design year, say ten years into the future. We may have information about the growth rate to be expected in this ten-year period for the whole study area; alternatively, we may have information on the likely growth in the number of trips originating and/or attracted to each zone. Depending on this information, we may be able to use different growth-factor methods in our estimation of future trip patterns.

4.1.3.1.1 Uniform Growth Factor

If the only information available is about a general growth rate τ for the whole of the study area, then we can only assume that it will apply to each cell in the matrix:

$$T_{ij} = \tau t_{ij}$$

where $\tau = T/t$, that is, the ratio of expanded over previous total number of trips.

Example 2: Consider the simple four-by-four base-year trip matrix of Table 4.4. If the growth in traffic in the study area is expected to be 20% in the next three years, it is a simple matter to multiply all cell values by 1.2 to obtain a new matrix as presented in Table 4.5.

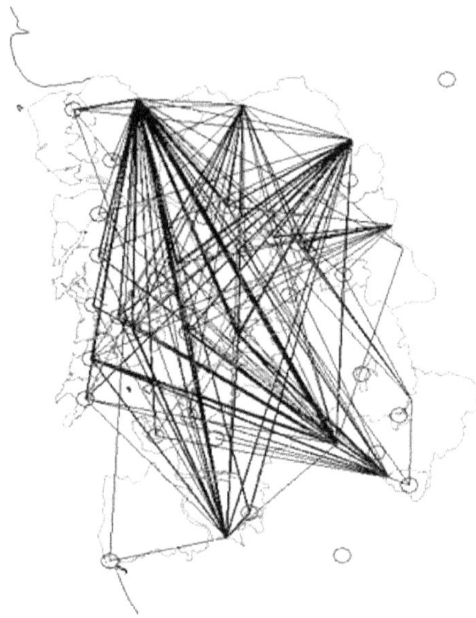

FIGURE 4.6 Desire Line Diagram of MMR

TABLE 4.4
Base-Year Trip Matrix

	1	2	3	4	Σ_j
1	5	50	100	200	355
2	50	5	100	300	455
3	50	100	5	100	255
4	100	200	250	20	570
Σ_i	205	355	455	620	1635

TABLE 4.5
Future Estimated Trip Matrix with $\tau = 1.2$

	1	2	3	4	Σ_j
1	6	60	120	240	426
2	60	6	120	360	546
3	60	120	6	120	306
4	120	240	300	24	684
Σ_i	246	426	546	744	1962

4.1.3.1.2 Singly Constrained Growth Factor

Consider the situation where information is available on the expected growth in trips originating in each zone, for example, shopping trips. In this case, it would be possible to apply this origin-specific growth factor (τ_i) to the corresponding rows in the trip matrix. The same approach can be followed if the information is available for trips attracted to each zone; in this case, the destination-specific growth factors (τ_i) would be applied to the corresponding columns. This can be written as follows:

Example 3: Consider the following revised version of Table 4.6 with growth predicted for origins:

This problem can be solved immediately by multiplying each row by the ratio of target over the base year total (Σ_j), thus giving the results in Table 4.7.

4.1.3.1.3 Doubly Constrained Growth Factors

An interesting problem is generated when information is available on the future number of trips originating and terminating in each zone say τ_i and T_j. This implies different growth rates for trips in and out of each zone and consequently having two sets of growth factors for each zone, say $F_{ij} = 0.5(\tau_i - T_j)$. The application of an average growth factor, say, is only a poor compromise as none of the two targets or

TABLE 4.6
Origin-Constrained Growth Trip Table

	1	2	3	4	Σ_j	Target O_i
1	5	50	100	200	355	400
2	50	5	100	300	455	460
3	50	100	5	100	255	400
4	100	200	250	20	570	702
Σ_i	205	355	455	620	1635	1962

TABLE 4.7
Expanded Origin-Constrained Growth Trip Table

	1	2	3	4	Σ_j	Target O_i
1	5.6	56.3	112.7	225.4	400	400
2	50.5	5.1	101.1	303.3	460	460
3	78.4	156.9	7.8	156.9	400	400
4	123.2	246.3	307.9	24.6	702	702
Σ_i	257.7	464.6	529.5	701.2	1962	1962

Source: Ortuzar and Willumsen (1999)

trip-end constraints will be satisfied. Historically a number of iterative methods have been proposed to obtain an estimated trip matrix which satisfies both sets of trip-end constraints, or the two sets of growth factors, which is the same thing.

All these methods involve calculating a set of intermediate correction coefficients which are then applied to cell entries in each row or column as appropriate. After applying these corrections to, say, each row, the totals for each column are calculated and compared with the target values. If the differences are significant, new correction coefficients are calculated and applied as necessary. The best known of these methods is due to Furness (1965), who introduced balancing factors A_i and B_j as follows:

$$T_{ij} = t_{ij}\tau_i \Gamma_j A_i B_j$$

or incorporating the growth rates into new variables a_i and b_j:

$$T_{ij} = t_{ij} a_i b_j$$

With $a_i = \tau_i A_i$ and $b_j = \Gamma_j B_j$.

The factors a_i and b_j (or A_i and B_j) must be calculated so that the constraints are satisfied. This is achieved in an iterative process which in outline is as follows:

1. Set all $b_j = 1.0$ and solve for a_j.
2. With the latest a_j solve for b_j, for example, satisfy the trip-attraction constraint.
3. Keeping the $b_j s$ fixed, solve for a_j and repeat steps 2 and 3 until the changes are sufficiently small.

This method produces solutions within 3–5% of the target values in a few iterations. There is not much point in enforcing the constraints to a level greater than the accuracy of the estimated trip-end totals. This method is often called a bi-proportional algorithm because of the nature of the corrections involved. This method is a special case of entropy maximizing models of the gravity type, if the effect of distance or separation between zones is excluded. In any case, the Furness method tries to produce the minimum corrections to the base-year matrix necessary to satisfy the future year trip-end constraints.

The most important condition required for the convergence of this method is that the growth rates produce target values T_i and $T - j$ such that

$$\sum_i \tau_i \sum_j t_{ij} = \sum_i T_j \sum_i T_{ij} = \mathrm{T}$$

Example 4: Table 4.8 represents a doubly constrained growth-factor problem. The solution to this problem (Table 4.9), after three iterations on rows and columns (three sets of corrections for all rows and three for all columns), can be seen in the table:

Note that this estimated matrix is within 1% of meeting the target trip ends, more than enough accuracy for this problem.

TABLE 4.8
Doubly Constrained Matrix Expansion Problem

	1	2	3	4	Σ_j	Target O_i
1	5	50	100	200	355	400
2	50	5	100	300	455	460
3	50	100	5	100	255	400
4	100	200	250	20	570	702
Σ_i	205	355	455	620	1635	
Target D_j	260	400	500	802		1962

TABLE 4.9
Solution to the Doubly Constrained Matrix Expansion Problem

	1	2	3	4	Σ_j	Target O_i
1	5.25	44.12	98.24	254.25	401.85	400
2	45.30	3.81	84.78	329.11	462.99	460
3	77.04	129.50	7.21	186.58	400.34	400
4	132.41	222.57	309.77	32.07	696.82	702
Σ_i	260.00	400.00	500.00	802.00	1962	
Target D_j	260	400	500	802		1962

Source: Ortuzar and Willumsen (1999)

4.1.3.1.3.1 Advantages and Limitations of Growth-Factor Methods Growth-factor methods are simple to understand and make direct use of observed trip matrices and forecasts of trip-end growth. They preserve the observations as much as is consistent with the information available on growth rates. This advantage is also their limitation as they are probably only reasonable for short-term planning horizons.

Growth-factor methods require the same database as synthetic methods, namely an observed (sampled) trip matrix; this is an expensive data item. The methods are heavily dependent on the accuracy of the base-year trip matrix. This is never very high for individual cell entries and therefore the resulting matrices are no more reliable than the sampled or observed ones. Any error in the base year may well be amplified by the application of successive correction factors. Moreover, if parts of the base-year matrix are unobserved, they will remain so in the forecasts. Therefore, these methods cannot be used to fill in observed cells of partially observed trip matrices.

Another limitation is that the methods do not take into account changes in transport costs due to improvements (or new congestion) in the network. Therefore they are of limited use in the analysis of policy options involving new modes, new links, pricing policies, and new zones.

4.1.3.2 Synthetic or Gravity Models

4.1.3.2.1 The Gravity Distribution Model

Distribution models of a different kind have been developed to assist in forecasting future trip patterns when important changes in the network take place. They start from assumptions about group trip making behavior and the way this is influenced by external factors such as total trip ends and distance traveled. The best known of these models is the gravity model, originally generated from an analogy with Newton's gravitational law. They estimate trips for each cell in the matrix without directly using the observed trip pattern; therefore, they are sometimes called synthetic as opposed to growth-factor models.

Probably the first rigorous use of a gravity model was by Casey (1955), who suggested such an approach to synthesize shopping trips and catchment areas between towns in a region. In its simplest formulation the model has the following functional form:

$$T_{ij} = \frac{\alpha P_i P_j}{(d_{ij})^2}$$

where P_i and P_j are the populations of the towns of origin and destination, d_{ij} is the distance between i and j, and α is a proportionality factor.

This was soon considered to be too simplistic an analogy with the gravitational law and early improvements included the use of total trip ends (O_i and D_j) instead of total populations, and a parameter n for calibration as the power for d_{ij}. This new parameter was not restricted to being an integer and different studies estimated value between 0.6 and 3.5. The model was further generalized by assuming that the effect of distance or separation could be modeled better by a decreasing function, to be specified, of the distance or travel cost between the zones. This can be written as:

$$T_{ij} = \alpha O D_j f(C_{ij})$$

where $f(C_{ij})$ is a generalized function of the travel costs with one or more parameters for calibration. This function often receives the name of deterrence function because it represents the disincentive to travel as distance (time) or cost increases. Popular versions for this function are as follows:

$$f(C_{ij}) = \exp(-\beta C_{ij}) \text{ exponential function}$$

$$f(C_{ij}) = C_{ij}^{n} \text{ power function}$$

$$f(C_{ij}) = C_{ij}^{n} \exp(-\beta C_{ij}) \text{ combines function}$$

4.1.3.2.1.1 Singly and Doubly Constrained Models The need to ensure that the restrictions are met requires replacing the single proportionality factor α by two sets of balancing factors A_i and B_j in the furnace model, yielding:

$$T_{ij} = A_i O_i B_j D_j f(C_{ij})$$

In a similar vein, one can again subsume O_i and D_j into these factors and rewrite the model as:

$$T_{ij} = a_i b_j f(C_{ij})$$

The above expression is a doubly constrained gravity model. Singly constrained versions, either origin or destination constrained, can be produced by making one set of balancing factors A_i and B_j equal to one. For an origin-constrained model, B_j is equal to 1 and for all j, and

$$A_i = 1 / \sum_j D_j f(C_{ij})$$

In this case of the doubly constrained model, the values of the balancing factors are:

$$A_i = 1 / \sum_j B_j D_j f(C_{ij})$$

$$B_j = 1 / \sum_i A_i O_i f(C_{ij})$$

The balancing factors are, therefore, interdependent; this means that the calculation of one set requires the values of the other set. This suggests an iterative process analogous to Furness which works well in practice: given set of values for the deterrence function $f(C_{ij})$, start with all $B_j = 1$, solve for A_i and then use these values to estimate the $B_j s$ repeat until convergence is achieved.

4.1.3.2.2 Calibrating a Gravity Model

Calibration of the gravity model is accomplished by developing friction factors and developing socioeconomic adjustment factors. As noted before, friction factors reflect the travel time of impedance on trip making. A trial-and-error adjustment process is generally adopted. One way, of course, is to use the factors from a past study in a similar urban area. Three items are used as input to the gravity model for calibration: (1) Production-attraction trip table for each purpose. (2) Travel times for all zone pairs, including intrazonal times. (3) Initial friction factors for each increment of travel time. Essentially, then, the calibration process involves adjusting the friction factor parameter until the planner is satisfied that the model adequately reproduces the trip distribution as represented by the input trip table—until the model's trip table agrees substantially with the table from the survey data, using indications such as the trip-time frequency distribution and the average trip time. The process is as follows: (1) Use the gravity model to distribute trips based on initial inputs. (2) Total trip attractions at all zones j, as calculated by the model, are compared to those obtained from the input observed trip table. (3) If this comparison shows significant differences, the attraction is adjusted for each zone, where a difference is observed. (4) The model is rerun until the calculated and observed attractions are reasonably balanced. (5) The models trip table and the input travel time table can be used for two comparisons: the trip-time frequency distribution and the average trip time. If there are significant differences, the process begins again.

4.1.4 Modal Split

The third stage in travel demand modeling is modal split. The trip matrix of O–D matrix obtained from the trip distribution is sliced into several matrices representing each mode. The choice of mode of transport is probably one of the most important classic models in transport planning. This is because of the key role played by public transport in policymaking. Public transport modes make more efficient use of road space than private transport. Also, they have more social benefits like if more people begin to use public transport, there will be less congestion on the roads and the accidents will be less. Again, in public transport, we can travel with low cost. In addition, the fuel is used more efficiently. The main characteristics of public transport are that they will have some schedule, frequency, etc.

On the other hand, private transport is highly flexible. It provides more comfortable and convenient travel. It has better accessibility also. The issue of mode choice, therefore, is probably the single most important element in transport planning and policymaking. It affects the general efficiency with which we can travel in urban areas. It is important then to develop and use models which are sensitive to those travel attributes that influence individual choices of mode.

4.1.4.1 Factors Influencing the Choice of Mode

The factors may be listed under three groups:

1. *Characteristics of the Trip Maker:* The following features are found to be important:
   a. car availability and/or ownership
   b. possession of a driving license
   c. household structure (young couple, couples with children, retired people, etc.)
   d. income
   e. decisions made elsewhere, for example, the need to use a car at work, take children to school, etc.
   f. residential activity

2. *Characteristics of the Journey:* Mode choice is strongly influenced by:
   a. the purpose of the trip; for example, the journey to work is normally easier to undertake by public transport than other journeys because of its regularity and the adjustment possible in the long run
   b. time of the day when the journey is undertaken
   c. late trips are more difficult to accommodate by public transport

3. *Characteristics of the Transport Facility:* There are two types of factors. One is quantitative and the other is qualitative.
   Quantitative factors are:
   a. relative travel time: in-vehicle, waiting, and walking times by each mode
   b. relative monetary costs (fares, fuel, and direct costs)
   c. availability and cost of parking

   Qualitative factors which are less easy to measure are:
   a. comfort and convenience
   b. reliability and regularity
   c. protection and security

4.1.4.2 Trip-End Modal Split Models

Traditionally, the objective of transportation planning was to forecast the growth in demand for car trips so that the investment could be planned to meet the demand. When personal characteristics were thought to be the most important determinants of mode choice, attempts were made to apply modal-split models immediately after trip generation. Such a model is called trip-end modal split model. In this way different characteristics of the person could be preserved and used to estimate modal split. The modal split models of this time related the choice of mode only to features like income, residential density, and car ownership.

The advantage is that these models could be very accurate in the short run, if public transport is available and there is little congestion. The limitation is that they are insensitive to policy decisions, for example, improving public transport, restricting parking, etc. would have no effect on modal split according to these trip-end models.

4.1.4.3 Trip-Interchange Modal Split Models

This is the post-distribution model; that is, modal split is applied after the distribution stage. This has the advantage that it is possible to include the characteristics of the journey and that of the alternative modes available to undertake them. It is possible to include policy decisions. This is beneficial for long-term modeling.

4.1.4.4 Aggregate and Disaggregate Models

Mode choice could also be aggregated if they are based on zonal and interzonal information. They can be called disaggregate if they are based on household or individual data.

4.1.4.5 Utility and Disutility Functions

A utility function measures the degree of satisfaction that people derive from their choices. A disutility function represents the generalized cost (akin to the concept of impedance) that is associated with each choice. The magnitude of either depends on the characteristics of each choice and on the characteristics of the individual making that choice. In the case of modal choice, the characteristics of the trip also bear a relationship to the utility associated with choosing a particular mode of travel. To specify a utility function, it is necessary to select both the relevant variables from this list and the particular functional form relating to the selected variables.

The disutility, in transportation, can be represented as

$$C_{ij} = a_1 t^{v}_{ij} + a_2 t^{w}_{ij} + a_3 t^{t}_{ij} + a_4 t_{nij} + a_5 F_{ij} + a_6 \phi_j + \delta$$

where t^{v}_{ij} is the in-vehicle travel time between i and j, t^{w}_{ij} is the walking time to and from stops, t^{t}_{ij} is the waiting time at stops, F_{ij} is the fare charged to travel between i and j, ϕ_j is the parking cost, and δ is a parameter representing comfort and convenience. If the travel cost is low, then that mode has more probability of being chosen.

4.1.4.6 Binary Logit Model

Binary logit model is the simplest form of mode choice, where the travel choice between two modes is made. Let there be two modes (m = 1, 2), then the proportion of trips by mode 1 from zone i to zone j is p^1_{ij}. Let c^1_{ij} be the cost of traveling from zone i to zone j using mode 1, and c^2_{ij} be the cost of traveling from i to zone j using the mode 2. There are three cases:

1. If $c^2_{ij} - c^1_{ij}$ is positive, then mode 1 is chosen.
2. If $c^2_{ij} - c^1_{ij}$ is negative, then mode 2 is chosen.
3. If $c^2_{ij} - c^1_{ij}$ is 0, then both modes have equal probability.

This relationship is normally expressed by a logit curve. Therefore, the proportion of trips by mode 1 is given by

$$p^1_{ij} = T^1_{ij}/T_{ij} = \frac{e^{-\beta c^1_{ij}}}{e^{-\beta c^1_{ij}} + e^{-\beta c^2_{ij}}}$$

This functional form is called logit, where c_{ij} is called the generalized cost and is the β parameter for calibration.

Example 5: Let the number of trips from zone i to zone j be 5000, and two modes are available which have the characteristics as shown in the table below. Compute the trips made by mode bus, and the fare that is collected from the mode bus. If the fare of the bus is reduced to 6, then find the fare collected.

	t_{ii}^{u}	t_{ii}^{w}	t_{ii}^{t}	f_{ij}	δ
Car	20	-	-	18	4
Bus	30	5	3	9	
a_i	0.03	0.04	0.06	0.1	0.1

Solution:

$$\text{Cost of travel by car} = c_{\text{car}} = 0.03*20+18*0.1+4*0.1 = 2.8$$

$$\text{Cost of travel by bus} = c_{bus} = 0.03*30+0.04*5+0.06*3+0.1*9 = 2.18$$

$$\text{Probability of choosing mode car} = p_{ij}^{car} = \frac{e^{-2.8}}{e^{-2.8}+e^{-2.8}} = 0.35$$

$$\text{Probability of choosing mode bus} = p_{ij}^{bus} = \frac{e^{-2.8}}{e^{-2.8}+e^{-2.8}} = 0.65$$

$$\text{Proportion of trips by car} = T_{ij}^{car} = 5000*0.35 = 1750$$

$$\text{Proportion of trips by bus} = T_{ij}^{bus} = 5000*0.65 = 3250$$

$$\text{Fare collected from bus} = T_{ij}^{bus}F_{ij} = 3250*9 = 29250$$

When the fare of bus gets reduce to 6,

$$\text{Cost of travel by bus} = c_{bus} = 0.03*30+0.04*5+0.06*3+0.1*6 = 1.88$$

$$\text{Probability of choosing mode bus} = p_{ij}^{bus} = \frac{e^{-1.88}}{e^{-2.8}+e^{-1.88}} = 0.175$$

$$\text{Proportion of trips by bus} = T_{ij}^{bus} = 5000*0.715 = 3575$$

$$\text{Fare collected from bus} = T_{ij}^{bus}F_{ij} = 3575*6 = 21450$$

Multinomial logit model

The binary model can be easily extended to multiple modes. The equation for such a model can be written as:

$$p_{ij}^{1} = \frac{e^{-\beta_{cij}^{1}}}{\sum e^{-\beta_{cij}^{1}}}$$

Example 6: Let the number of trips from i to j be 5000, and three modes are available which have the following characteristics. Compute the trips made by the three modes and the fare required to travel by each mode.

	t_{ii}^{u}	t_{ii}^{w}	t_{ii}^{t}	f_{ij}	δ
Car	20	-	-	18	4
Bus	30	5	3	6	-
Train	12	10	2	4	-
a_i	0.03	0.04	0.06	0.1	0.1

Solution:

$$\text{Cost of travel by car} = c_{car} = 0.03*20+18*0.1+4*0.1 = 2.8$$

$$\text{Cost of travel by bus} = c_{bus} = 0.03*30+0.04*5+0.06*3+0.1*6 = 1.88$$

$$\text{Cost of travel by train} = c_{train} = 0.03*12+0.04*10+0.06*2+0.1*4 = 1.28$$

$$\text{Probability of choosing mode car} = p_{ij}^{car} = \frac{e^{-2.8}}{e^{-2.8}+e^{-1.88}+e^{-1.28}} = 0.1237$$

$$\text{Probability of choosing mode bus} = p_{ij}^{bus} = \frac{e^{-2.18}}{e^{-2.8}+e^{-1.88}+e^{-1.28}} = 0.3105$$

$$\text{Probability of choosing mode train} = p_{ij}^{train} = \frac{e^{-2.18}}{e^{-2.8}+e^{-1.88}+e^{-1.28}} = 0.5657$$

$$\text{Proportion of trips by car} = T_{ij}^{car} = 5000*0.1237 = 618.5$$

$$\text{Proportion of trips by bus} = T_{ij}^{bus} = 5000*0.3105 = 1552.5$$

$$\text{Proportion of trips by train} = T_{ij}^{train} = 5000*0.5657 = 2828.5$$

$$\text{Fare collected from bus} = T_{ij}^{bus} * F_{ij} = 1552.5*6 = 9315$$

$$\text{Fare collected from train} = T_{ij}^{train} * F_{ij} = 2828.5*4 = 11314$$

4.1.4.7 The Nested Logit Model

Figures 4.8 and 4.9 compare the multinomial logit (MNL) and a nested logit structure involving three mode choices: the automobile, a local bus service, and rail transit (Figure 4.7). The MNL places these modes on a single level. By contrast, the nested structure groups the bus and rail together as sub-choices of the composite transit mode. This structure permits a change in the utility of one of the transit modes to affect the share of the other transit mode to a greater degree than a mode that does not belong to the transit nest. In other words, a greater degree of choice substitution is allowed within nests than between nests.

Examining the top-level decision of whether to travel by automobile (A) or transit (T) gives

$$P(A) = \frac{\exp(U_A)}{\exp(U_A + \exp(U_T))} \quad \text{and} \quad P(T) = \frac{\exp(U_T)}{\exp(U_A + \exp(U_T))}$$

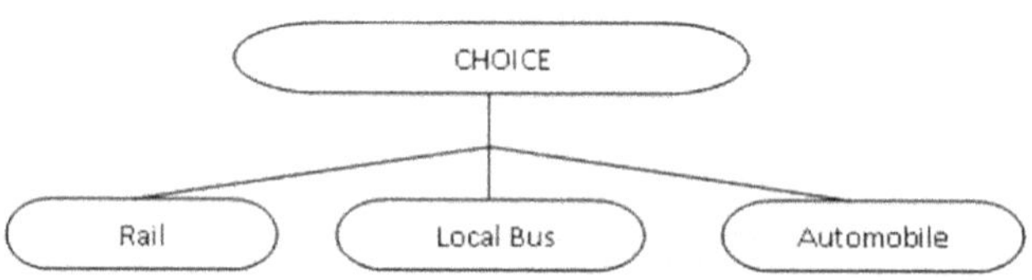

FIGURE 4.7 Multinomial Logit Structure for a Three-Mode Choice

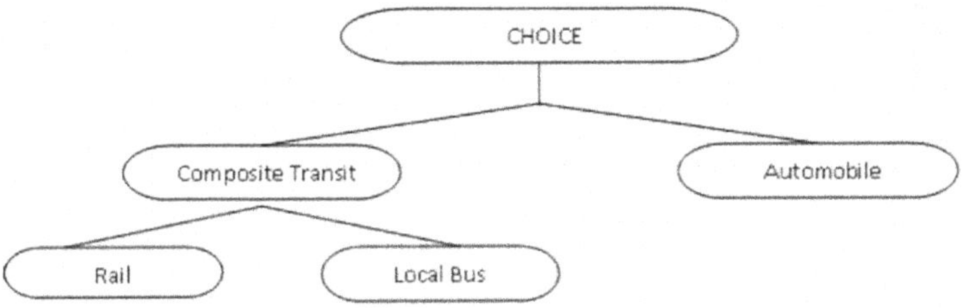

FIGURE 4.8 Nested Logit Structure

where the composite transit utility $U_T = f(U_B, U_R)$.

By moving to the lower transit level, the conditional probabilities of choosing the bus (B) or the rail (R), given the decision to travel by transit, become

$$P(B/T) = \frac{\exp(U_B)}{\exp(U_B + \exp(U_R))} \quad \text{and} \quad P(R/T) = \frac{\exp(U_R)}{\exp(U_B + \exp(U_R))}$$

To calculate the unconditional probabilities of choosing bus or rail, use the following equations:

$$PB = P(B/T) * P(T) \quad \text{and} \quad P(R) = P(R/T) * P(T).$$

The utility of the composite transit mode needs to capture the characteristics of all transit submodes (i.e., bus and rail). This is normally accomplished by including in the transit utility expression the *Logsum* variable multiplied by its calibration coefficient:

$$Logsum = \ln \exp(U_B) + \exp(U_R).$$

The numerical value of the *Logsum* coefficient resulting from estimating the model provides information about the appropriateness of the selected nesting structure. If the estimated value of $\theta = 0$, the transit utility is independent of the utilities of the submodes. Consequently, the primary choice between transit and auto is not affected by changes in the utilities of the submodes. Any such change redistributes the market shares of the submodes solely between them. In this case, the submodes are said to be perfect substitutes of each other.

If the estimated value of turns out to be greater than 0 but less than 1, the selected structure is acceptable. A value of exactly equal to 1 implies that there exists an equivalent MNL model that is equally appropriate, whereas a value greater than 1 indicates that the selected nesting structure is inappropriate, and other structures need to be investigated.

Example 7: An estimation procedure for a mode choice model of the nested logit structure shown in Figure 6.8 found that $U_T = a_t + \theta * logsum$ with $a_t = -0.42$

and $\theta = 1.0$. For a particular zonal interchange, the following modal utilities were calculated in accordance with the estimated nested logit model:

$$U_A = -0.36, U_B = -0.88, U_R = -0.78$$

Calculate (1) the corresponding mode shares, and (2) the effect of a policy that is expected to cause a change $\Delta U_R = -0.10$.

Solution:

Part 1: Baseline conditions

Nest Level

Mode m	*Um*	*exp(Um)*	*p(m / T)*
B	−0.88	0.415	0.475
R	−0.78	0.458	0.525
		$\Sigma = 0.873$	$\Sigma = 1.00$

$$U_T = -0.52 + 1.0 - ln(0.839) = -0.56$$

Primary choice level

Mode m	*Um*	*exp(Um)*	*p(m / T)*
A	−0.36	0.698	0.550
T	−0.56	0.571	0.450
		$\Sigma = 1.366$	$\Sigma = 1.000$

$$P(B) = P(B/T) * P(T) = 0.475 * 0.450 = 0.214$$

$$P(R) = P(R/T) * P(T) = 0.525 * 0.450 = 0.236$$

$$P(A) = 0.450$$

Part 2: After change
Nest Level

Mode m	*Um*	*exp(Um)*	*p(m / T)*
B	−0.88	0.415	0.500
R	−0.88	0.415	0.500
		$\Sigma = 0.830$	$\Sigma = 1.000$

$$U_T = -0.42 + 1.0 * ln(0.830) = -0.61$$

Primary choice level

Mode m	*Um*	*exp(Um)*	*p(m / T)*
A	−0.36	0.698	0.462
T	−0.61	0.543	0.438
		Σ = 1.241	Σ = 1.000

$$P(B) * P(B/T) * P(T) = 0.438 * 0.500 = 0.219$$

$$P(R) * P(R/T) * P(T) = 0.438 * 0.500 = 0.219$$

$$P(A) = 0.562$$

4.1.5 Trip Assignment

The last phase of the four-step transportation planning process is concerned with the trip-makers' choice of path between pairs of zones by travel mode and with the resulting vehicular flows on the multimodal transportation network. This step may be viewed as the equilibration model between the demand for travel and the supply of transportation in terms of the physical facilities and, in the case of the various possible mass transit modes, the frequency of service provided. Following are the methods employed for the purpose of trip assignment.

4.1.5.1 Free/All-or-Nothing Traffic Assignment

The free/all-or-nothing assignment technique allocates the entire volume interchanging between pairs of zones to the minimum path calculated on the basis of free-flow link impedance. After all interchange volumes are assigned, the flow on a particular link is computed by summing all interzonal flows that happen to include that link on their minimum paths.

Example 8: Assign the following interzonal vehicular trips emanating from zone 1 to the network shown.

J	2	3	4	5
Qij	800	500	200	600

The minimum tree emanating from zone 1 is reproduced in Figure 4.11. The interzonal flows using each link of the tree are assumed to compute the total contribution of the given flows to these links. Thus, link 7–2 takes only the total interchange between zones 1 and 2, and link 7–8 takes the flow from zone 1 to zone 3. Link 6–7 takes the sum of the flows from zone 1 to zone 2 and from zone 1 to zone 3 because it belongs to both minimum paths. These links may also be assigned additional flows if they happen to be part of minimum paths that originate from zones other than zone 1.

4.1.5.2 Free/Multipath Traffic Assignment

In essence, a free/all-or-nothing assignment assumes that all trip makers traveling between a specific pair of zones actually select the same path. Interchange volumes

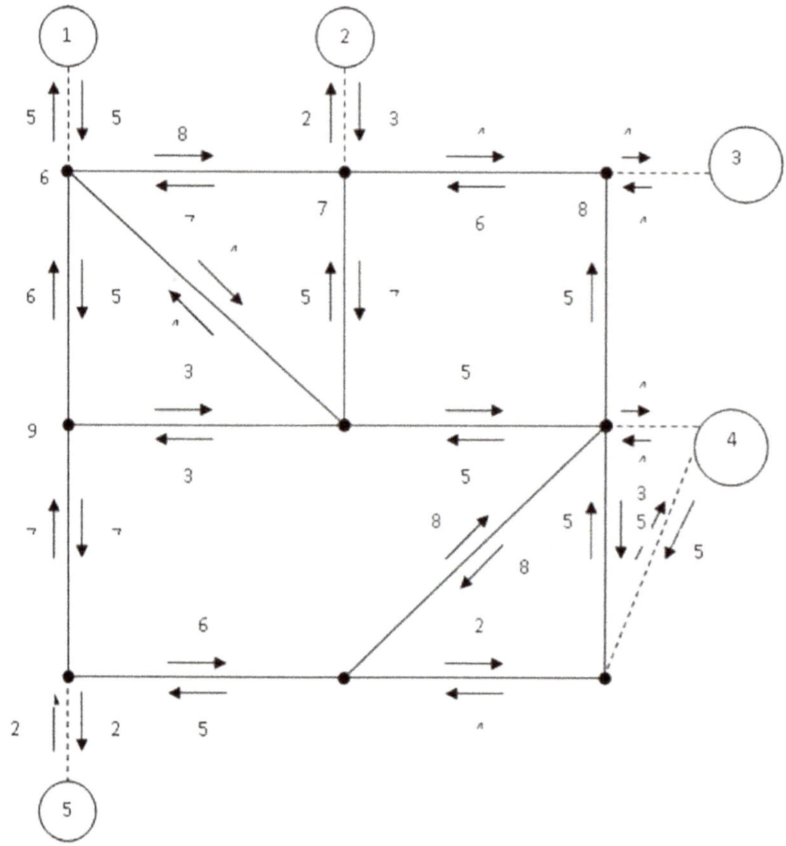

FIGURE 4.9 Hypothetical Network

are divided among a number of paths, and algorithms that are capable of determining several paths between each pair of zones in order of increasing impedance are available. Therefore, it is possible to approve the interchange volume between these paths according to some realistic rule. Irwin and von Cube suggested the following inverse-proportion function to compute the fraction to be assigned to each of several interzonal routes:

$$P(r) = \frac{W_{IJr}^{-1}}{\sum_x \left(W_{IJx}^{-1}\right)}$$

where W_{IJr} is the impedance of route r from I to J. As the following example illustrates, the use of the MNL model with disutilities based on path impedance is another possibility.

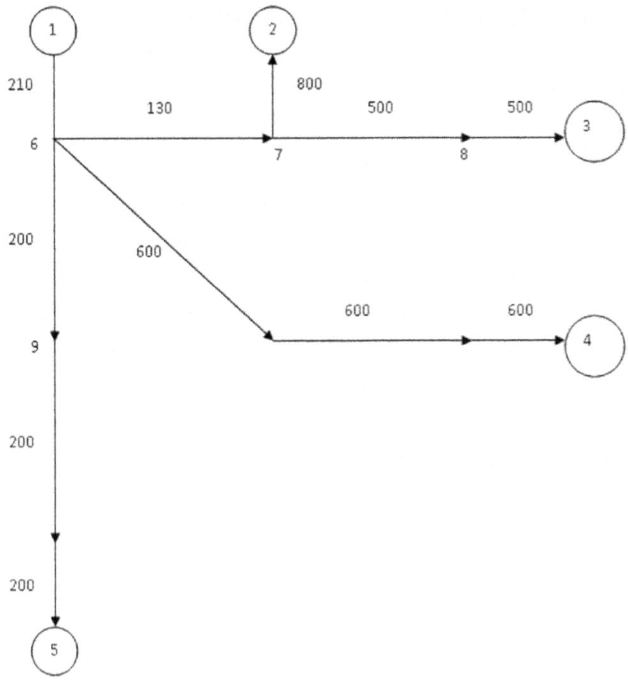

FIGURE 4.10 Minimum Tree

FIGURE 4.11 Peak Hour Traffic Loading on Transport Network of Bengaluru

4.1.5.3 Capacity-Restrained Traffic Assignment

As the flow increases toward capacity, the average stream speed decreases from the free-flow speed to the speed at maximum flow. Beyond this point, the internal friction between vehicles in the stream becomes severe, the traffic conditions worsen, and severe shock waves and slow-moving platoons develop.

The implication of this phenomenon on the results of free-traffic assignment presents the following paradox: The interzonal flows are assigned to the minimum paths computed on the basis of free-flow link impedances. But if the link flows were at the levels dictated by the assignment, the link speeds would be lower, and the link travel times would be higher than those corresponding prior to trip assignment may not be the minimum paths after the trips are assigned. Several iterative assignment techniques address the convergence between the link impedances assumed prior to assignment and the link impedances that are implied by the resulting link volumes. These techniques are known as capacity-restrained methods or techniques that employ capacity restraints. The relationship between link flow and link impedance is described as the link-capacity function. Several such functions are found in technical literature. One of the forms, developed by BPR, is expressed mathematically as

$$w = \overline{w}\left[1+0.15\left(q / q_{max}\right)^{4}\right]$$

where w is the impedance of a given link at flow q, $\overline{w}$ is the free flow impedance of the link, $qmax$ is the link's capacity, and q is the link flow.

This function states that at capacity the links impedance is 15% higher than the free-flow impedance. If the demand were to exceed the capacity of the link, the resulting shock waves and their dissipation times would cause a rapid deterioration in the link flow conditions. The outcome of the whole four-stage travel demand modeling process is the assessment of traffic volume on each and every link of the integrated transport network and the assessment of congestion levels on each link in terms of volume to capacity (*v/c*) ratio. Figures 4.11 and 4.12 show the traffic assignment output for some case examples.

4.1.6 Use of Travel Demand Models

Travel demand modeling is the utilization of a computer software package to replicate the real-world transportation system around us (roads, intersections, traffic control devices, congestion delays, use of a transit system, etc.). Once the computer model can accurately replicate the existing conditions of a study area, it can then be used to predict future travel patterns and demands based on changes in the transportation system (e.g., new public transport system, new roads, wider roads with more capacity, closed roads, new bicycle lanes); changes in the land use (e.g., more residential development, a new industrial site); and changing demographics (more or less people in a specific area, access to a vehicle, etc.). By simulating the current multimodal transport network conditions and the travel demand on those networks, deficiencies in the system can be identified. It is also an important tool in planning future network enhancements and analyzing currently proposed projects that may

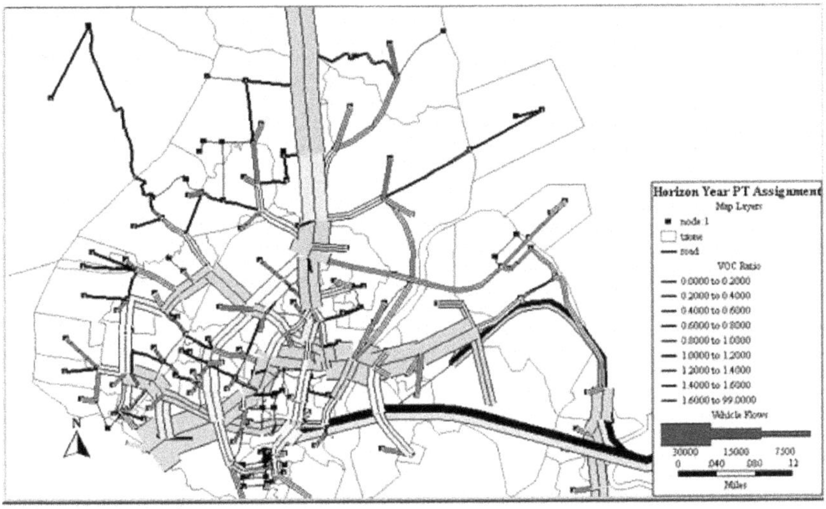

FIGURE 4.12 Peak Hour Traffic Loading on Transport Network of Thane

involve a combination of different modes of transport. Travel demand models are developed to simulate actual travel patterns and existing demand conditions.

Networks are constructed using current network inventory files for each mode containing data for each link within the network. Travel demand is generated using socioeconomic data such as household size, population density, income, automobile availability, employment data, etc. Once the existing conditions are evaluated and adjusted to satisfactorily replicate actual travel patterns, the model inputs are then altered to project future-year conditions. Using these inputs, the model is able to derive future capacity limitations relative to the current transportation system that may consist of different modes. Once these deficiencies are identified, potential improvements are evaluated by rerunning the model with an improved or modified transportation system. A range of different mode split levels, transport networks, and even different land-use patterns are tested this way. For example, Table 4.10 gives desired modal share for Indian cities based on consideration for sustainability, etc. This can then be taken as a basis to arrive at various combinations of different mode infrastructure mix, which can be tested using developed travel demand models. Future-year traffic projections are based on numerous assumptions about how population, employment, automobile operating costs, public transport cost, and other factors will change over time. As such, future-year projections are only as good as the assumptions that are made.

Travel demand modeling remains a cornerstone of the integrated transportation infrastructure planning process. Travel demand models have achieved sophistication unimaginable by those who laid their foundation in the 1950s. The current generation of travel demand models, however, is unable to analyze many transportation policies under consideration in urban areas, such as pricing, demand management, and air quality strategies. Modern travel demand models can predict highway volumes

TABLE 4.10
Desired Modal Share for Indian Cities

City Population (in Million)	Mass Transport	Bicycle	Other Modes
0.1–0.5	30–40	30–40	25–35
0.5–1.0	40–50	25–35	20–30
1.0–2.0	50–60	20–30	15–25
2.0–5.0	60–70	15–25	10–20
5.0+	70–85	15–20	10–15

reasonably well in many circumstances, but they lack precision in several critical areas. These areas include prediction of latent and induced travel demand, intermodal impacts, and land-use/transportation relationships. Widening highways and developing new highway corridors in built-up urban areas are often impractical and always expensive. Shortly after suffering the cost and disruption of adding highway capacity, communities see congested conditions return long before the end of the useful life of the facility and well in advance of travel-model horizon years. Although much of this congestion can be attributed to induced demand that is, to travelers shifting from unimproved to improved routes, is also the result of realized latent demand that is, to travel resulting from a decline in the cost of travel. We understand this phenomenon only in terms of observed trends but not in terms of the underlying behavioral forces that cause it to occur. Models capable of predicting latent demand are necessary to properly simulate the effect of transportation policies over prolonged periods.

4.1.7 Induced Travel Demand

While we propose different infrastructure projects, it is important to understand that the benefits projected are sometimes inflated due to lack of consideration of latent and induced demand, which surfaces as a result of change in mode choice and route choice behavior of travelers in light of proposed infrastructure improvements. For example, if a road is widened from say two lanes to four lanes to improve the travel time of traffic using that road, the resultant improvements may not be as anticipated since it is not the same traffic that will continue to use it but more traffic may come on this road either because some people now find it more attractive or because some people leave public transport and use private vehicle on this road since they find it more attractive to travel by their own modes using this widened road now. Similar phenomena can occur if, say, capacity of a metro rail line is doubled. So, induced demand, or latent demand, is the phenomenon that after supply increases more of a good is consumed. This is entirely consistent with the economic theory of supply and demand; however, this idea has become important in the debate over the expansion of transportation systems and is often used as an argument against widening roads, such as major commuter roads. It is considered by some to be a contributing factor to urban sprawl.

Traffic engineers often compare traffic to fluid, assuming that a certain volume must flow through the road system. However, urban traffic may be more comparable to a gas that expands to fill available space. Road improvements that reduce travel costs attract trips from other routes, times and modes, and encourage longer and more frequent travel. This is called generated traffic, referring to additional vehicle traffic on a particular road. This consists in part of induced travel, which refers to increased total vehicle miles travel (VMT) compared with what would otherwise occur. Generated traffic reflects the economic law of demand, which states that consumption of good increases as its price declines. Roadway improvements that alleviate congestion reduce the generalized cost of driving (i.e., the price), which encourages more vehicle use. Put another way, most urban roads have latent travel demand, additional peak-period vehicle trips that will occur if congestion is relieved. In the short run, the transport generated represents a shift along the demand curve; reduced congestion makes driving cheaper per mile or kilometer in terms of travel time and vehicle operating costs. Over the long run, induced travel represents an outward shift in the demand curve as transport systems and land-use patterns become more automobile dependent, so people must drive more to maintain a given level of accessibility to goods, services, and activities. Accurate transport planning and project appraisal must consider these three impacts:

1. Generated traffic reduces the predicted congestion reduction benefits of road capacity expansion.
2. Induced travel imposes costs, including downstream congestion, accidents, parking costs, pollution, and other environmental impacts.
3. The additional travel that is generated provides relatively modest user benefits, since it consists of marginal value trips (travel that consumers are most willing to forego).

Ignoring these factors distorts planning decisions. Experts conclude that the economic value of a scheme can be overestimated by the omission of even a small amount of induced traffic. This matter is of profound importance to the value-for-money assessment for the road improvement programs as quite small absolute changes in traffic volumes have a significant impact on the benefit measures. The induced travel effects of changes in land use and trip distribution may be critical to an accurate evaluation of transit and highway alternatives (Litman, 2011).

Having the background of travel demand modeling, as discussed in the previous section, it is easy to take into account the induced demand effect due to the provision of a new facility or any change in the existing infrastructure. This can be done by reflecting the changes in the mode choice and/or network files and rerunning the traffic assignment with revised mode share and network to obtain revised flows on the network, which will eventually reflect the changes in flow due to latent and induced demand effect. Further, we can test various combinations of mode integration and their corresponding infrastructure provisions so as to not only overcome the negative effects of induced demand but also shift it in favor of more sustainable modes of transport, for example, inducing or generating (latent) more demand in favor of public transport by improving cycling and walking facilities to access public transport and thus providing better door-to-door connectivity compared to travel by private modes.

4.1.8 Case Study—Bengaluru (Tunnel Road and Double-Decker Roads)

Bengaluru, often referred to as India's Silicon Valley, is facing a severe transportation crisis as its population is projected to reach 30 million in the near future. Traffic congestion, along with rising air and noise pollution, has placed immense strain on the city's infrastructure, necessitating urgent interventions. To address these challenges, sustainable and integrated mobility solutions, such as metro and suburban rail systems, have been proposed to enhance connectivity and reduce congestion. However, alongside these transit initiatives, the government has also put forward infrastructure projects such as double-decker roads and a North–South tunnel corridor, which require careful evaluation regarding their effectiveness and long-term implications.

The Bengaluru Suburban Rail Project (BSRP) is a major initiative aimed at improving urban mobility by seamlessly integrating with the BMRCL system. While projects like the BSRP and metro expansion focus on sustainable solutions, double-decker roads and tunnel corridors raise concerns about their potential to promote private vehicle dependency rather than encouraging a shift toward public transport. These high-cost interventions often result in limited long-term benefits, as they primarily serve private automobiles while failing to provide substantial relief from congestion.

The proposed double-decker roads are conceptualized to optimize space utilization along key metro corridors. These include the Outer Ring Road West Line, Magadi Road Metrolite, Hebbal–Sarjapur Metrolite, and additional sections in Phase 3 of the metro expansion. Additionally, a 3.2 km corridor along the Yellow Line has already been constructed. While these structures are intended to ease congestion, their actual impact on public transportation usage remains questionable. The introduction of elevated roads risks shifting travel demand away from metro corridors, reducing ridership on sustainable modes and leading to inefficiencies in the overall transport network.

Similarly, the proposed tunnel corridor, connecting Hebbal in the north to the Central Silk Board in the south, aims to reduce travel times by providing a dedicated underground roadway. While such a system may offer initial relief to surface level congestion, it inherently encourages car ownership and usage, counteracting the benefits of metro and suburban rail investments. Tunnel roads also present significant environmental and safety risks, including vulnerability to flooding and increased emissions due to higher traffic volumes.

The methodology employed in this study is shown in Figure 4.13. First, the daily ridership is estimated, followed by forming the scenarios for year 2027, 2031, and 2041 for metro-only, suburban-only, and Mass Rapid Transit System (MRTS) scenarios. Another two scenarios are formed for year 2041 which are double-decker scenario and double-decker plus tunnel road scenario. After these scenarios are formed, relevant digitization of proposed metro and suburban corridors is done using QGIS and TransCAD. A conventional four-stage travel demand model (TDM) is used to obtain demand estimation for all the formed scenarios. This travel demand model is validated for base year 2022. After trip assignment in passenger per hour per direction (pphpd) in the relevant networks for all the scenarios, corridor-wise demand is

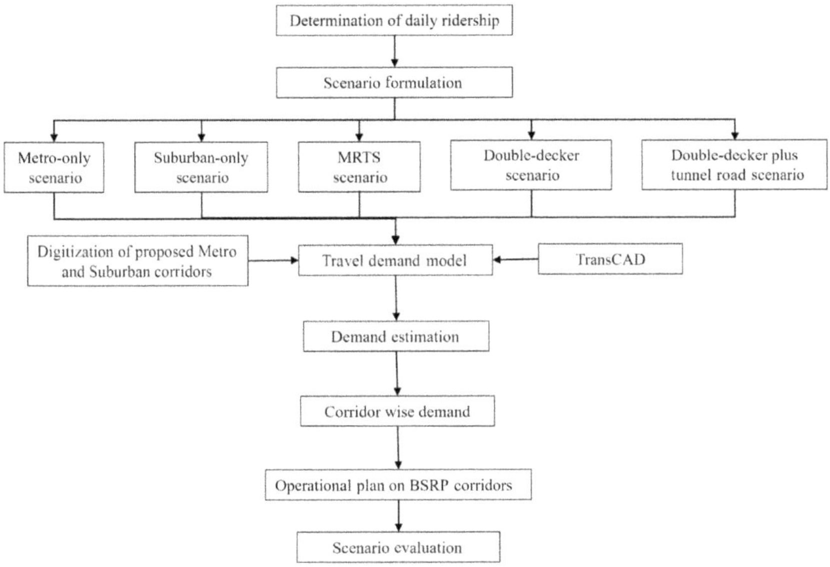

FIGURE 4.13 Research Methodology

obtained for metro and suburban sections along with details about the maximum section loadings.

The proposed construction of double-decker corridors (flyover cum metro) lies along the following metro corridors as per the data provided by BMRCL (refer Figures 4.14 and 4.15).

1. Corridor 1 from JP Nagar Fourth Phase to Hebbal of 29.20 km length of Phase 3, along the ORR West Line Corridor of Metro
2. Corridor 2 from Hosahalli to Kadabagare of 11.45 km length of Phase 3, along the Magadi Road Metrolite.
3. Corridor 3 from Sarjapur to Ibbalur (ORR Junction) of 14 km length of Phase 3, along the Hebbal–Sarjapur Metrolite.
4. Corridor 4 from Agara to Koramangala Third Block of 2.45 km length in Phase 3-A corridor, along the Hebbal–Sarjapur Metrolite.

Additionally, a fifth corridor of 3.2 km from Raggigudda to Silk Board along the Yellow line of the Bengaluru Metro Rail Project has already been constructed.

The proposed Tunnel Road connects the Hebbal in the North to the Central Silk Board in the South and is expected to reduce travel time considerably. Figure 4.16 shows the location of the Tunnel Road.

Figure 4.16 shows that there are moderate ridership changes, with losses and gains in ridership on the introduction of tunnel roads and double-decker roads. Analysis of projected scenarios for 2041 indicates that the integration of double-decker roads and tunnel corridors could lead to notable shifts in ridership patterns. Without seamless

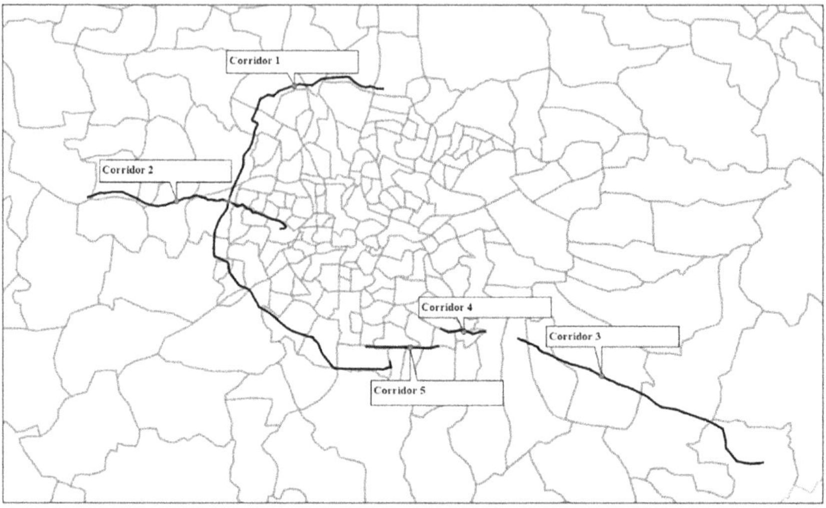

FIGURE 4.14 Double-Decker Corridors

FIGURE 4.15 Double-Decker Corridors

connectivity with existing transit systems, these road-based solutions may cause up to a significant reduction in metro and suburban rail ridership. Additionally, while some corridors, such as the ORR West Line, may experience minor ridership gains, overall public transport usage is likely to decline. The preference for private vehicles, fueled by these projects, exacerbates congestion at entry and exit points, increasing travel delays and environmental degradation.

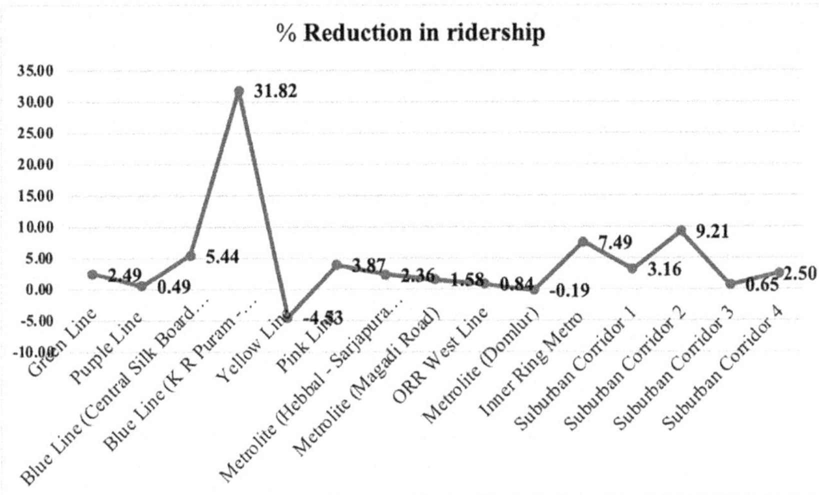

FIGURE 4.16 Percentage Reduction in Ridership for Double-Decker and Tunnel Road Scenario (MRTS)

In contrast, a well-integrated MRTS has the potential to achieve an 80% modal share for sustainable transportation by 2041. Strategic planning, including intermodal connectivity and transit-oriented development, is essential to ensure public transit remains the backbone of Bengaluru's urban mobility. Investments should prioritize expanding the metro and suburban rail services with scalable capacity enhancements, such as double-tracked corridors and increased train frequencies, rather than promoting infrastructure that reinforces car dependency.

Ultimately, while double-decker roads and tunnel corridors appear to offer immediate solutions to traffic congestion, their long-term impact may be detrimental to Bengaluru's sustainability goals. These projects divert focus from public transport investments and encourage urban sprawl, exacerbating the very challenges they seek to solve. A more effective approach would be to strengthen metro and suburban rail networks, ensuring that all areas within the BMR have equitable access to reliable and efficient public transport.

4.2 CURRENT TRANSPORT INFRASTRUCTURE AND GAP IDENTIFICATION WITH RESPECT TO SHORT-, MEDIUM-, AND LONG-TERM ESTIMATED DEMAND PATTERNS

When planning PMT in urban areas, it is important to understand both the current transport infrastructure and the future travel needs of the population. Transport planning is typically divided into three categories based on the time frame: short-term, medium-term, and long-term planning. Each type of plan addresses different needs and involves varying levels of investment and complexity.

4.2.1 Short-Term Planning

Short-term plans focus on immediate solutions and have a time horizon of 5 years or less. These plans are designed to quickly address specific transport challenges, such as reducing traffic congestion or resolving bottlenecks. They often involve low-cost measures and practical adjustments, such as introducing traffic signals, creating one-way streets, or implementing "tidal flow" systems (where traffic direction changes during peak hours). The goal of short-term planning is to provide quick relief with minimal financial and structural investment.

4.2.2 Medium-Term Planning

Medium-term plans are designed for a period of about 10 years and require moderate investment in engineering improvements. These plans aim to address issues that need more substantial changes than what short-term plans can provide. Examples of medium-term solutions include building pedestrian subways, providing off-street parking facilities, or upgrading existing infrastructure. These initiatives focus on improving safety, accessibility, and traffic flow in a sustainable manner.

4.2.3 Long-Term Planning

Long-term plans, also called comprehensive or strategic plans, look ahead to a time horizon of around 20 years. These plans aim to create a well-structured vision for the future of urban transport systems. They involve large-scale projects and significant investments, often forming the backbone of Master Plans for cities. Long-term planning includes the development of new transport infrastructure, such as metro rail systems, expressways, or multimodal hubs, with the goal of supporting future urban growth and meeting the evolving mobility needs of the population.

For a transportation network to work efficiently and fairly, it must accommodate various travel needs through a multimodal approach. This means offering diverse options such as walking, cycling, public transit, and private vehicles to cater to all types of users. When non-motorized modes like walking and biking, or public transit systems, are underdeveloped or neglected, several challenges arise:

- Parents may need to drive children to nearby schools due to a lack of safe sidewalks or bike paths.
- Inadequate public transit options might force commuters to rely on personal vehicles, even if they would prefer other modes of travel.
- Households may be burdened with owning more vehicles than they can afford because their community lacks accessible alternatives.

Historically, many transportation policies and plans have favored car travel, often neglecting affordable and resource-efficient modes. This has also led to urban sprawl, where cities grow outward in a car-dependent manner rather than focusing on compact, multimodal development. To address these challenges, transport planning needs to prioritize and invest in walking, cycling, and public transit options while aligning with broader goals such as reducing costs, protecting the environment, and improving public health.

Before the mid-twentieth century, walking, cycling, and public transit were widely recognized as essential modes of transportation. However, over the last century, transportation planning became heavily car focused. As a result, road systems were developed extensively to support convenient car travel, while other modes were largely ignored. Today, while driving is often convenient and safe, non-automobile travel options are frequently inadequate. This has created gaps in the transport system, such as:

- Insufficient infrastructure for pedestrians and cyclists.
- Limited and unreliable public transit services.
- Unsafe and inaccessible options for people who do not or cannot drive.

This imbalance has led to unmet travel demands and forced many people to rely on cars, even when they would prefer or benefit from alternative modes.

To address the diverse needs of all travelers, transport planning must take a broader approach. This involves considering the following:

1. *Current Users of Non-Automobile Modes:* Those who already rely on walking, biking, or public transit.
2. *Occasional Users:* People who might use these modes occasionally if better infrastructure or services were available.
3. *Potential Users (Latent Demand):* Individuals who would shift to walking, cycling, or transit if these modes were made safer and more convenient.
4. *Community-Wide Benefits:* Advantages for everyone when fewer people drive, such as reduced congestion, lower emissions, and less pressure on parking.
5. *Strategic Objectives:* Goals like affordability, better mobility for non-drivers, and reduced environmental impact.

To develop a comprehensive transportation plan, the following steps are typically involved:

1. *Assess Current Conditions:* Evaluate the performance of the existing transport network.
2. *Predict Future Growth:* Estimate population and job growth to identify key development areas.
3. *Identify Challenges and Needs:* Understand current and future transportation issues and propose potential solutions.
4. *Analyze and Prioritize Solutions:* Evaluate different strategies and projects to determine which are most effective.
5. *Draft Plans and Programs:* Develop both long-term and short-term plans with specific actions and projects.
6. *Secure Funding:* Create a financial plan to implement the selected initiatives.

Adopting these steps with a focus on multimodal transport will help create an inclusive, efficient, and sustainable transportation system that reduces car dependence while improving mobility for everyone.

A study by Verma et al. (2024) analyzes the proposed transportation solutions for the BMR, focusing on sustainable and integrated mobility systems to tackle severe traffic congestion and pollution. Using a four-stage travel demand model validated for the base year 2022, the study explores scenarios for the milestone years 2027, 2031, and 2041. It compares metro-only, suburban rail-only, and integrated MRTS configurations. The findings highlight the significant advantages of a well-integrated MRTS network, which is projected to achieve an 80% modal share for sustainable transportation by 2041, while reducing congestion and improving urban connectivity. The study also assesses the impact of unsustainable infrastructure projects like double-decker roads and tunnel corridors. These projects, while offering localized travel time reductions, encourage private vehicle use, increase pollution, and exacerbate environmental and social inequalities. Additionally, such interventions fail to address the root causes of congestion and result in higher vehicle dependency.

The study underscores the importance of enhancing suburban rail capacity, ensuring seamless integration of metro and suburban rail systems, and providing comprehensive MRTS coverage across the BMR to promote equitable and scalable urban mobility. Recommendations include designing MRTS corridors with double tracks from the outset to minimize future disruptions and accommodate growing demand. The findings also stress that double-decker roads and tunnel corridors prioritize private vehicles over public transit and non-motorized modes, resulting in lower person-carrying capacity compared to metro or light rail systems. Furthermore, tunnel roads are vulnerable to flooding during extreme weather events, posing safety risks and compromising urban resilience. Policymakers are urged to prioritize sustainable transit solutions that reduce congestion, pollution, and urban inequalities, while aligning infrastructure investments with long-term mobility, environmental, and social goals. This comprehensive approach offers valuable insights into building a resilient and inclusive urban transport framework for Bengaluru.

4.3 ROUTING AND SCHEDULING OF PMT SERVICES

Many techniques have been used to solve the basic problem of routing and scheduling in public transportation. Unfortunately, no technique has incorporated all or even most of the practical options or real-life restrictions and extensions. Different versions of the vehicle routing and scheduling problems on the transportation network appear in all fields of transportation depending on the specific objective. Well-organized vehicle routes or well-designed schedules can remarkably contribute toward a decrease in transportation costs and increase the quality of the transportation services. For the operation of public buses, users and operator are the two agencies involved, both having conflicting objectives to be satisfied. Thus, the transit network design problems become complex due to the following:

1. Difficulty in combining user costs and operator costs in a single objective function.
2. Difficulty in optimizing the objective function.
3. Discrete nature of route design and various other constraints involved with route coverage, route duplication, route length, and directness of service.

Various attempts have been made to solve the routing and scheduling problem with multiple objective functions. Much remains to be done using available computationally efficient algorithms.

Routing and scheduling are planning issues in a transit network, which can be grouped into the following three major categories:

1. *Perform Maximum Transportation Work:* This is expressed by the number of passenger-trips or passenger-km. This implies the provision of high travel speed, passenger convenience, and other elements that attract passengers. In planning of a given network, if average trip lengths do not vary much among alternative designs, these two objectives may have similar values; however, in comparing an urban transit system with a regional one, both passenger-trips and passenger-km values should be considered because they may be quite different alternative designs.
2. *Achieve Maximum Operating Efficiency:* This objective can ultimately be expressed as the minimum total system cost for a required performance level.
3. *Creative Positive Impacts:* These include a variety of effects, from short-range ones, such as reduction of highway congestion, to the long-range goals, such as achievements of high population mobility, desirable land-use patterns, sustainability, and high quality of life.

These three categories of objectives generally correspond, respectively, to the requirements of the three interested parties considered in transit system evaluation: passengers, operators, and community.

4.3.1 Passenger Attraction

The greater the number of passenger trips a transit system carries, the better it serves citizens' and a city's needs. The more persons-km a transit network carries, the more economically it operates and the more it reduces less socially desirable automatic travel. Consequently, the dominant goal in designing most transit systems is to attract as many passengers as possible. The main network design features that affect transit service quality and thereby influence passenger attraction are defined here.

1. Area coverage, defined as percentage of the total urban area that is within 5-minute and 10-minute walking of transit stations, represents the basic element of transit system availability. For planning purposes, it can be considered that most potential transit users within a 5 minute (400 m) walking distance

will use the system. Between that distance and a 10 minute (800 m) walking distance, the potential ridership gradually decreases practically to zero.

2. Operating speed is mostly a function of ROW category and station spacings. In street operations (ROW category C), the speed depends on street and traffic conditions; in rapid transit (ROW category A), station spacing is usually the determination factor. Station planning is therefore a major element influencing travel speeds.
3. Transit lines should be designed to follow as closely as possible major origin–destination (O–D) patterns of travel in order to attract and serve efficiently the maximum number of passengers.
4. Direction of travel, measured as the ratio of passenger-km traveled on the network to passenger-km along straight O–D lines, expresses an efficiency of the network design. For a given travel demand, the network configuration that results in the smallest ratio is most efficient.
5. Simplicity, connectivity, and easy transfers must be considered in design because they are important for passenger convenience and attraction. The greatest operational simplicity is achieved by operating lines independently of each other with easy passenger transfer among them.

4.3.2 Network Operating Efficiency

Operating efficiency, representing performance and cost of transit operations, is a major concern of the second interested party.

There are five major determinants of efficiency.

1. Continuity and balancing of lines, or provisions of direct services among areas with heavy travel demand, is desirable for both passenger convenience and operating efficiency. Continuity should be planned so that each line has reasonably well-balanced capacity requirements on all of its major sections.
2. Operating flexibility is increased through provisions of connections between lines at which TUs can be switched among lines for scheduled or unforeseen change of service. Greater extensiveness of networks and track connections among lines always increase operating flexibility. Turn back tracks, usually center dead-end tracks connected with both through tracks; allow short-turn operations, that is, use of intermediate stations as terminals.
3. Integration with other modes ensures that a network provides a service in an area greater than immediate sheds of its stations. It allows efficient utilization of joint rail/highway corridors. Most important are intermodal stations, which permit easy interchange of passengers between different modes such as bus feeders to rail trunk lines, P + R facilities, etc.
4. Terminals, depots, and yards should be placed at such locations that deadheading, that is, travel between depots and the lines, is minimized. Also because these facilities occupy large areas, they should be located where cost of land is not high. Both requirements make outlying sites better suited for depot and yard locations than central city areas. Terminals should be placed near major streets to allow good accessibility by walking and vehicles of different access modes.

5. Cost of the system is, of course, often the most important single factor in network design. Investment cost depends mostly on ROW category and alignment, its type (at grade, elevated, tunnel, or other), and stations (their size, complexity, equipment, storage facilities, etc.).

4.3.3 Network City Interactions

To ensure that the transit system will have positive impacts and minimal undesirable effects on the community (the third-interesting party), the following major factors must be considered in its design.

1. Land use should always be a major factor in planning a transit network, particularly metro and other rail systems. In many cities, the main objectives of building a metro system are to enable and stimulate development of certain land-use patterns.
2. Topography and environment usually impose stringent requirements and constraints on a transit system. Hilly terrain restricts the possibilities of the network design. The existence of valleys, rivers, or lakes often gives natural corridors for transit facilities.
3. Major geographical constraints, such as hills, valleys, rivers, or bays requiring major tunnels or bridges for crossing, always make construction of high-capacity transit lines far more efficient than construction of roads because two-track transit lines have capacities equivalent to ten or more highway lines per direction.
4. The existing transportation network must be taken into account since it may be utilized for location of transit facilities (bus lanes, light rail medians, metro tunnels, or railway corridors), or it may sometimes create obstacles, such as difficult at-grade crossings.

As mentioned, the influence of most of these factors on transit network design cannot be defined precisely; however, it is important that the designer be familiar with them and considers them carefully.

4.3.4 Optimum TU Capacity

The following example referred from V.R. Vuchic (2005) explains how optimum TU capacity is obtained.

On a transit line with a given p_{max}, the operator is interested in providing a required transporting capacity by a smaller number of large TUs because of their lower cost per unit capacity. However, passengers require shorter headways, that is, and higher frequency of service, which dictates the use of a greater number of smaller TUs.

To find the optimal balance between the interests of the two parties considering costs only, a model including the operating costs $C_o(R_s / h)$ and passenger costs $C_p(R_s / h)$ should be developed and used to derive the expression for the total cost, $C = C_o + C_p$. With mathematical expressions for these costs, the TU capacity that results in the minimum total costs C should then result in the minimum total cost C. This is illustrated here by a rather simple model of a transit line with length L (km), headway h (minutes), cycle time T (minutes), and load factor a_{max} (prs/sp). The TU (bus or train) capacity is C_{TU} (sps/TU).

Simplifying assumptions for this model are:

1. Total hourly cost of operation of the line C_o is equal to the product of unit vehicle operating cost c_o (R_s/TU-h) and the number of TUs on the line NTU $C_o = c_o * \text{NTU}$.
2. Passenger demand (design hour volume p_{max}) is constant and is related to the total number of passengers boarding along the entire line, PL, by a coefficient: $p_{max} = \eta_p$ PL.
3. Cycle speed V_c on the line is independent of TU size and of the number of passengers.
4. Average passenger waiting time is equal to half of the headway, h/2.
5. Average value of passenger time is c_p (R_s/prs-h).

First the number of TUs is expressed by the given line parameters. Using the below equations one obtains the equations for T and h:

$$T = \frac{120L}{V_c} \text{ and } h = \frac{60\alpha_{max} * C_{TU}}{P_{max}}$$

$$\left|\frac{T,h}{min}\right|\frac{L}{km}\left|\frac{V}{km/h}\right|\frac{\alpha}{prs/sp}\left|\frac{C_{TU}}{sps/TU}\right|\frac{P}{prs/h}\right| \tag{4.1}$$

N_{TU} can be expressed by these operating elements as:

$$N_{TU} = \frac{T}{h} = \frac{2L \cdot P_{max}}{\alpha_{max} \cdot C_{TU} \cdot V_c}$$

$$\left|\frac{N}{TU}\right|\frac{T,h}{min}\left|\frac{L}{km}\right|\frac{V}{km/h}\left|\frac{\alpha}{prs/sp}\right|\frac{C_{TU}}{sps/TU}\left|\frac{P}{prs/h}\right| \tag{4.2}$$

Operating cost can now be expressed, using the first assumption above as:

$$C_o = N_{TU} \cdot C_o = \frac{2L \cdot P_{max}}{\alpha_{max} \cdot C_{TU} \cdot V_c} \cdot c_o$$

$$\left|\frac{C}{RS/h}\right|\frac{L}{km}\left|\frac{V}{km/h}\right|\frac{\alpha}{prs/sp}\left|\frac{C_{TU}}{sps/TU}\right|\frac{P}{prs/h}\left|\frac{c_o}{Rs/TU_h}\right. \tag{4.3}$$

Passenger cost for all passengers on the line, P_L is:

$$C_p = P_L \cdot \frac{h}{120} \cdot c_p = \frac{P_L \cdot \alpha_{max} \cdot C_{TU}}{2P_{max}} \cdot c_p$$

$$\left|\frac{C}{RS/h}\right|\frac{P}{prs/h}\left|\frac{h}{min}\right|\frac{c_p}{Rs/(prs-h)}\left|\frac{\alpha}{prs/sp}\right|\frac{C_{TU}}{sps/TU}\right| \tag{4.4}$$

Total cost, consisting of the operation and passenger cost, is obtained by summing up Equations 4.3 and 4.4:

$$C = \frac{2L \cdot P_{max}}{\alpha_{max} \cdot C_{TU} \cdot V_c} \cdot c_o + \frac{P_L \cdot \alpha_{max} \cdot C_{TU}}{2P_{max}} \cdot c_p$$
$$\left| \frac{C}{RS/h} \right| \frac{P}{prs/h} \left| \frac{h}{min} \right| \frac{c_p}{Rs/(prs-h)} \left| \frac{\alpha}{prs/sp} \right| \frac{C_{TU}}{sps/TU} \left| \frac{c_o}{Rs/TU_h} \right| \quad (4.5)$$

To find the optimum value of vehicle capacity that will minimize total cost, the derivative of Equation 4.5 with respect to C_{TU} is set equal to 0:

$$\frac{C}{C_{TU}} = -\frac{2L \cdot P_{max}}{\alpha_{max} \cdot C_{TU}^2 \cdot V_c} \cdot c_o + \frac{P_L \cdot \alpha_{max}}{2P_{max}} \cdot c_p \quad (4.6)$$

from which the optimum vehicle capacity, C_{TU}^*, is:

$$C_{TU}^* = \frac{2P_{max}}{\alpha_{max}} \sqrt{\frac{L \cdot c_o}{P_L \cdot V_c \cdot c_p}} \text{(units as above)} \quad (4.7)$$

Figure 4.17 shows a diagram of C_o, C_p, and C as a function of C_{TU}, as well as the position of C_{TU}^*

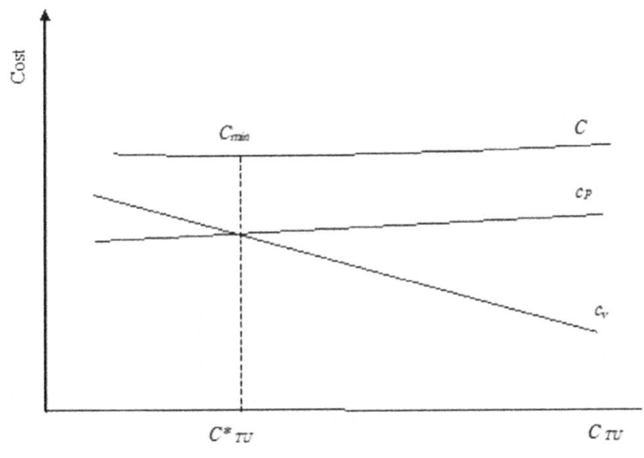

FIGURE 4.17 Operator and Passenger Costs as Functions of TU Capacity

4.4 STRATEGIES EXHIBITED BY GENERAL PUBLIC (DIFFERENT SECTIONS OF THE PUBLIC) WITH REGARD TO CHOICE OF MODE

The choice of transportation mode by individuals is influenced by a complex interplay of factors ranging from travel time and cost to perceptions of safety, comfort, and accessibility. These factors often vary across different demographic groups, cultural contexts, and trip purposes. By analyzing these elements, transportation planners can better understand the behavior of commuters and develop systems that meet their diverse needs effectively.

Many studies highlight travel time and cost are two of the most critical factors shaping transportation choices (Mayo and Tabaoda, 2020). For shorter trips, individuals often prefer modes that offer speed and convenience, such as ride-hailing services or personal vehicles. In contrast, longer trips typically require modes that provide reliability and comfort, such as trains or intercity buses. Cost also plays a significant role, particularly for lower-income groups who prioritize affordability over speed or luxury. For example, public transportation is often the go-to option for low-income households due to its cost-effectiveness, while wealthier individuals may opt for private vehicles or premium transit services. Balancing travel time and cost is essential for transit operators to appeal to a wide range of users.

The availability and accessibility of transportation options significantly influence mode selection. Proximity to transit hubs, parking facilities, and seamless connections between different modes play a pivotal role in determining a commuter's choice. Inadequate infrastructure or poorly designed transit systems can deter commuters from choosing public transportation, underscoring the need for well-planned and accessible networks.

Sociodemographic characteristics such as age, income, gender, and occupation also affect transportation decisions. Income levels often dictate the affordability of certain modes, with lower-income groups leaning toward public transit and higher-income individuals favoring private vehicles or faster, more expensive options. Age further influences preferences: older commuters are more likely to use public transit if the service quality is improved, while younger individuals prefer personal vehicles due to their higher mobility needs. Gender differences are also evident, with women in countries like India, Thailand, and Malaysia showing a greater preference for public transit, often driven by safety concerns. On the other hand, men in nations such as Indonesia and Japan are more inclined toward driving.

Safety and comfort are fundamental aspects influencing modal choices. Commuters prioritize systems that ensure personal safety, operational reliability, and comfort during travel. For example, women often prefer public transportation options with well-lit stations, secure environments, and reserved seating areas. Comfort-related factors, such as smooth rides, adequate seating, and low noise levels, are also essential for many commuters. People often place significant emphasis on ride comfort and convenience when making their choices. Systems that fail to meet safety and comfort expectations risk losing patronage, particularly from demographics with higher sensitivity to these issues.

Cultural norms, environmental consciousness, and habitual behaviors also play a significant role in mode selection. In some societies, such as India, car ownership is perceived as a status symbol, leading to a preference for private vehicles over public transit. Conversely, growing awareness of environmental issues has led many to adopt greener options such as bicycles, walking, or electric buses. Behavioral patterns further influence decisions; for instance, Filipino commuters prefer using jeepneys for short, affordable trips due to familiarity but avoid them for longer journeys, citing discomfort and safety concerns.

The purpose of a trip significantly influences modal choice. For work commutes, reliability and efficiency are paramount, with commuters favoring modes that minimize delays and offer direct routes. For leisure or social trips, flexibility and convenience become more critical, leading to a preference for ride-hailing services or personal vehicles. School travel often prioritizes safety and reliability, with parents opting for school buses or secure public transit options. Understanding these variations in trip purpose can help transit providers tailor their services to meet specific needs.

The strategies employed by the general public in choosing transportation modes are influenced by a myriad of factors, including sociodemographics, travel attributes, cultural norms, and contextual considerations. Understanding these factors is essential for designing transportation systems that cater to diverse needs. Improving travel time, cost, accessibility, safety, and comfort while addressing first- and last-mile connectivity challenges is key to increasing public transit usage. By integrating these insights into policy and planning, transit systems can become more inclusive, efficient, and environmentally sustainable.

4.5 LOS CONCEPT IN PLANNING URBAN MOBILITY

As technology in the automobile sector has advanced, efforts have also been made to upgrade public transport systems to match these innovations. However, public transport still faces challenges in competing with the private car. To help close this gap, it is crucial to have a precise and consistent way to measure public transport service. This kind of system allows for the management and continuous improvement of existing modes of transport while also defining clear targets for future improvements. The concept of LOS is essential when evaluating public transport systems. It involves assessing various factors that contribute to the user experience, such as speed, density, travel time, comfort, and convenience. Though each mode of transport has specific physical characteristics, these common features are crucial in determining how passengers perceive the overall quality of service. By using these characteristics as benchmarks, planners can understand what needs to be improved to meet passengers' needs.

Botzow et al. identify several key factors for evaluating the LOS in public transport. Travel time is crucial, as it directly correlates with the system's average speed. High-speed modes like trains may reach 80 kph, while buses in congested areas average only 6 kph. Headway, or the time between vehicle departures, must be short enough to encourage use; ideally, it should not exceed 5 minutes to improve efficiency and user satisfaction. Transfers add to the total travel time and inconvenience, so minimizing them is essential. Each transfer typically adds a 5-minute penalty to the overall journey. Fare collection systems that delay passengers should be streamlined with automation to improve efficiency, particularly when delays exceed 30 seconds. Passenger comfort, affected by factors like crowding, temperature, and noise, is vital for user satisfaction. Comfort should avoid extreme unpleasant but tolerable conditions may deter passengers. Finally, design and operational features such as infrastructure and safety standards influence overall service levels and must balance cost with efficiency.

Efficient first- and last-mile connectivity is also essential for improving the LOS in public transit systems. These connections refer to the ease with which passengers can access the transit system at the start of their journey (first mile) and reach their final destination after completing their trip (last mile). When these connections are poorly designed or inefficient, they can significantly impact on the overall travel time, often making the entire trip longer than the core transit segment itself. For example, passengers may have to walk long distances to access the nearest transit station or rely on limited or inconvenient transfer options, such as buses or bikes, to complete the journey. In some cases, the first- and last-mile segments can exceed the travel time of the main trip, creating delays that discourage people from using public transit.

This inefficiency reduces the overall LOS and can lead to decreased patronage. As travel times become longer and less predictable, passengers are more likely to turn to alternative modes of transport, such as private vehicles or taxis, which may offer more direct and time-efficient routes. This shift not only lowers the ridership but also undermines efforts to promote sustainable and efficient urban mobility.

4.6 STAKEHOLDER CONSULTATION AND PREFERENCES WITH RESPECT TO FIRST-MILE AND LAST-MILE CONNECTIVITY

A key part of the planning process is ensuring that people can easily make the first and last parts of their journeys, which often cover short distances. For example, many passengers need a quick way to get from their home to the nearest metro station (first mile) or from a bus stop to their office (last mile). If this part of the trip is inconvenient, people may prefer to use private vehicles, which leads to more traffic congestion and pollution.

One very important step in addressing the first- and last-mile connectivity needs of the commuters is to first understand the needs and requirements of people. This can be done in multiple ways. For example, in a city such as Bengaluru which is divided in around 225 wards, planners can group and combine 15–20 wards and hold stakeholder meetings in these agglomerations. These stakeholder workshops or meetings should be held with people from all walks of life for uniform representativeness. This can be followed by understanding the needs of first- and last-mile connectivity of each of these groups in terms of walking needs, feeder buses, shared autos, and parking facilities.

Another way to do this is through a survey that asks people about their short-distance travel habits. This survey can find out how many trips are short in distance, where these trips begin and end, and which routes they follow most often. By having clear information on the number of such trips and their destinations, planners can understand the importance of first- and last-mile travel in the overall transport system. With this data in hand, authorities can design short-distance services that fit people's needs. For instance, they could create special routes for small vehicles, such as minibuses or shared auto-rickshaws, that run frequently and have fixed fares. These routes could be auctioned to private operators under government guidelines. Because these trips often follow predictable patterns, having fixed routes and set fares would make it easier and more affordable for people to reach mass transport stations or stops. This would not only save passengers' time and money but also give them a more reliable and comfortable experience.

Improving short-distance travel in this way can boost the satisfaction of passengers who use mass transit, such as metros and buses. When passengers find it easy to reach a station or get to their final destination after using mass transit, they are more likely to use these systems regularly. Over time, this will increase the overall popularity of public transport. In the long run, making first- and last-mile travel more convenient, predictable, and efficient helps create a stronger, more attractive, and sustainable urban transport network. This approach encourages people to rely more on public transportation, reduces the number of private vehicles on the roads, and supports better mobility for everyone in the city.

Many public authorities seek to incorporate public participation in their decision-making processes, including those processes involving transportation policy and planning. This is done for a number of reasons. At its best, the public participation process is designed to help assure that the people's will is done (Innes and Booher, 2000, p6) in an attempt to design things to meet the aspirations of its citizens. In some cases, transportation planners seek to involve the community while the planning and decision-making are still readily influenced and attempt to take the opinions and values of the community into account when developing the plan (Shepherd and Bowler, 1997).

Presumably, along with the desire to do the people's will, there are two general assumptions that underline the decision to include some public participation in decision-making (Mumpower, 2001):

1 the decision-making process can be improved by the inclusion of knowledge and perspectives that would otherwise be absent; and
2. if the public has better access to knowledge and is more able to influence the process, then it may develop a sense of project ownership and thus increase its support.

Open houses, or town hall meetings, are a time-honored method of gauging public opinion regarding proposed transportation projects. However, such processes are highly self-selective because those people who are motivated in some way show up, whereas most others do not. So, the question arises: to what extent are the opinions expressed in an open house a valid indication of actual public opinion? This section considers the issue of measuring public opinions by discussing a study of public opinions regarding a proposed transportation master plan for the City of Edmonton. The same survey was conducted with both open-hour attendees and a randomly selected sample of the population. The instrument asked the respondent in each case to indicate preferences regarding the potential directions that the master plan could take by allocating points to alternative directions considered in sets of two or three possibilities. For example, the respondent was to allocate 12 points between improving accessibility for the mobility disadvantaged versus saving money (by reducing municipal expenditures on transportation). A given respondent would indicate a preference for a complete emphasis on improving accessibility at the expense of saving money by allocating all 12 points to the first of the two possibilities or indicate a preference for a balance between the two possibilities by allocating 6 points to each. Each respondent was asked to perform a total of 11 of these allocations involving other such potential directions as improving personal mobility by automobile, improving mobility for commerce, preserving and enhancing community integrity, improving personal mobility by bicycle, protecting the city's environment, protecting the global environment, using debt to finance transportation expenditure, and using municipal taxes to finance transportation expenditure. A total of 472 (self-selected) open-house attendees and 868 randomly selected individuals responded. Analysis of the results indicates that the two groups displayed significantly different patterns in their responses, with the open-house group displaying a greater degree of extremism where the allocations of points tended more to 0 and 12. Notwithstanding these differences, it was also the case that the two groups tended to indicate the same sorts of preference in general but with the open-house group showing stronger tendencies. This suggests that such indications of preference from open houses can be taken to be indicative of the directions but not necessarily the magnitudes of the corresponding preferences for the population more generally. Different methods used in this system are discussed in the following section.

4.6.1 The Edmonton Opportunity

In 1997, the City of Edmonton sought public opinions using both of these aforementioned methods as part of a larger survey. With both methods participants were asked to complete the same questionnaire. The collection of data from these two groups of individuals using the same instrument provided an opportunity for comparing the two methods in order to gauge the similarities and differences between these two samples and their corresponding populations.

The intent of the research reported here was to identify and examine any differences in the indications provided by the resulting two samples and, more formally, to establish if there is a significant bias in the self-selected sample obtained in the open-house process in particular.

4.6.2 Key Choices Survey

The Edmonton Key Choices Survey was conducted between April 25 and May 24, 1997 as part of the development of the Edmonton Transportation Master Plan.

The Transportation Master Plan was to become the general outline according to which the city would develop its transportation infrastructure and related transportation policy. As such, the Transportation Master Plan had the potential to influence significantly the future of the city and its citizens impacting everything from the mobility of goods and people to the nature of the urban form and the resulting quality of life provided.

The Key Choices Survey questionnaires were distributed in two ways. The city conducted a series of open-house meetings concerning the Transportation Master Plan, and all attendees were asked to complete the questionnaire as part of the meeting. Questionnaires were also mailed out to a randomly selected set of households in the City of Edmonton, together with materials describing the Transportation Master Plan process. These households were also subsequently contacted by telephone and encouraged to respond.

4.6.3 The Questionnaire

The questionnaire included a series of 11 worksheets designed to elicit the respondents' attitudes to elements of policy relevant to the Transportation Master Plan, along with some standard questions seeking personal information about the respondent. In each of these worksheets, the respondent was asked to allocate a set of 12 points among 2 or 3 different possible planning objectives to indicate the relative degrees of emphasis the respondent felt was appropriate among the objectives. For example, with two competing objectives, A and B, if the respondent felt A was all important relative to B, then A would get 12 points and B would get 0 points, whereas if the respondent felt A and B were equally important, then both A and B would get 6 points.

The first six worksheets asked the respondent to allocate 12 points among two alternatives each time, as follows:

1. improving mobility versus saving money;
2. improving accessibility for the mobility disadvantaged versus saving money;
3. improving mobility for commerce versus preserving and enhancing neighborhood integrity;
4. improving personal mobility by automobiles versus preserving and enhancing neighborhood integrity;
5. improving personal mobility by automobile versus protecting the city's environment; and
6. personal mobility by automobile versus the protecting the global environment.

The last five worksheets asked the respondent to allocate 12 points among three alternatives each time, as follows:

7. improving personal mobility versus protecting the environment versus saving money
8. improving personal mobility by automobile versus improving personal mobility by transit versus improving personal mobility by bicycle;

9. improving accessibility for the mobility disadvantaged, versus improving alternatives to auto use versus saving money;
10. protecting the global environment versus protecting the city's environment versus preserving and enhancing neighborhood integrity; and
11. using debt to finance transportation expenditure versus using user charges to finance transportation expenditure versus using municipal taxes to finance transportation expenditure.

In total, 472 completed questionnaires were obtained at the eight public open houses hosted by the city; and 868 completed questionnaires were returned by the mail-out sample group (representing a response rate of 29%).

The observations in the mail-out sample were then weighted so that the joint distribution for the sample matched that for the population (according to Statistics Canada tables) in terms of age, gender, household size, and dwelling type. All of the results for the mail-out sample presented here use this weighted version of the sample.

4.6.4 Results of the Abovementioned Methods

4.6.4.1 Distributions of Points

Figure 4.18 compares the distributions of points obtained for worksheet 1 for the two samples. Similar sorts of two-dimensional and three-dimensional distributions were obtained for the other two-alternative and three-alternative worksheets, and these are presented elsewhere (Mackay, 2004).

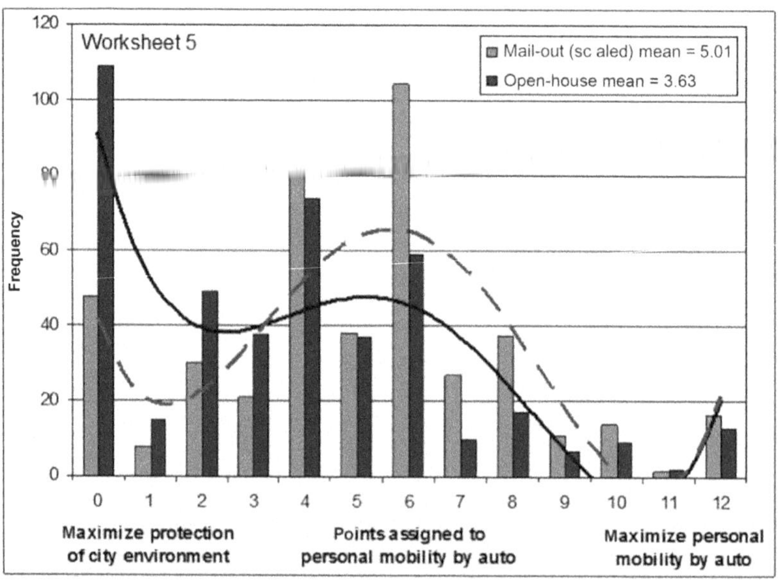

FIGURE 4.18 Distributions of Points Allocated in Worksheet

4.7 FEASIBILITY FOR HIGHER-ORDER MODES LIKE BRT AND METRO RAIL

4.7.1 Bus Rapid Transit

The term BRT is used for any bus line that has partial separation of lanes, new buses, or distinctive line designation. BRT must also have some or most of the following features.

1. Stations and stops that allow simultaneous stopping of two or more buses, their overtaking or platooning to provide operations with short headways, and a line capacity greater than 3000–5000 spaces per hour.
2. Engines with low-level pollution and noise.
3. Convenient transferring to other buses and rail lines.
4. Prepaid or contactless fare collection, which minimize delays during boarding and alighting.

4.7.2 High-Performance Transit Modes

The basic features for provision of high-performance transit service are operation on ROW category A only. With exclusive use of protected ROW, guided modes represent the logical technology because guidance allows use of high-capacity electric trains with signal control. One of these modes is RRT (Metro).

4.7.3 RRT (Metro)

RRT, popularly better known as the metro, utilizes high-capacity electric trains with high acceleration and braking rates. It represents the high-performance transit modes with the lowest operating cost per space-km. Its implementation requires very high investment and extensive construction, including disruption of areas along future lines. However, RRT systems have virtually unlimited life and exert a strong, permanent impact on mobility of population, urban form, and character.

4.7.4 Interdependence of ROW and System Technology

The basic relationship between investment costs of the three ROW categories and mode performance is shown graphically in Figure 4.19. The box for each row category shows generally the area where the performance-cost package (PCP) for the systems in that category belongs. ROW category C, street transit, requires a distinctly lower investment than medium- and high-performance modes using ROW categories B and A, respectively. The diagram shows the problem that occurs when a city uses only buses on streets and rapid transit: the former has a relatively low performance, while the latter requires a large investment. A large area of medium-performance modes at medium investment cost levels is provided mainly by LRT, BRT, and AGT modes with their relative PCP positions as shown.

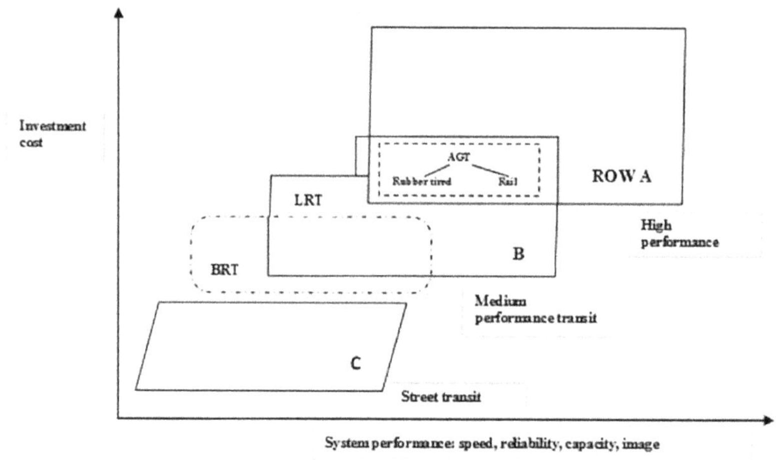

FIGURE 4.19 PCPs of Different Generic Classes of Transit Modes

4.7.5 Comparison of RRT and LRT

A comparison of RRT with its closest widely used neighbor in the family of modes, standard LRT, is given below.

The advantages of metro compared to LRT are as follows:

1. With ROW A only and full signal control, RRT has higher speed, reliability, and safety.
2. Due to long trains and rapid passenger exchange at stations, RRT has a much higher line capacity.
3. With its high performance and strong image, RRT has stronger passenger attraction.
4. Distinct image of its stations and system performance give RRT greater potential for positive impacts on urban development than any other transit mode.

The disadvantages are as follows:

1. RRT requires substantially higher investment cost, causes more disruption due to construction, and requires longer implementation time than LRT.
2. RRT requires more rigid alignment than LRT and cannot penetrate pedestrian areas.
3. LRT is more conducive to construction in stages, utilizing different ROW categories, while RRT is limited to ROW A only.

4.7.6 Light Rail Compared to Buses

Comparison of all major medium-performance modes is presented here. The first comparison, given in Tables 4.11–4.13, includes three modes: RB, BRT, and LRT.

TABLE 4.11
Comparison of RB, BRT, and LRT (System Components)

Characteristics/Mode	RB	BRT	LRT
ROW	C	B (C, A)	B (C, A)
Support	Road	Road	Rail
Guidance	Steered	Steered	Guided
Propulsion	ICE	ICE	Electric
TU control	Driver/visual	Driver/visual	Driver/visual
Vehicle capacity	80–120	80–180	100–250
Max TU size	Single vehicle	Single vehicle	1–4 car trains
Max TU capacity	120	180	4*180 = 720

Note: ICE: Internal Combustion Engine

TABLE 4.12
Comparison of RB, BRT, and LRT (Lines/Operational Elements)

Characteristics/Mode	RB	BRT	LRT
Lines	Many		
Headways on each line	Long/medium/short	Long/medium/short	Medium/short
Stop spacings (m)	80–250	200–400	250–600
Transfers	Few	Some/many	Many

TABLE 4.13
Comparison of RB, BRT, and LRT (System Characteristics)

Characteristics/Mode	RB	Bus Rapid Transit	LRT
Investment cost/km	Low	High	Very high
Operating cost/space	Medium	Medium	Low
System image	Variable	Good	Excellent
Passenger attraction	Limited	Good	Strong
Impacts on land use and city livability	Least	Moderate	Strongest

Their characteristics are grouped in three categories: system components, lines and operational components, and system characteristics.

4.7.7 Comparison of Main BRT and LRT Features

The evaluation and comparison of BRT and LRT are different in many aspects. Generally, BRT requires lower investment and shorter implementation time than LRT, but LRT provides higher service quality, attracts more passengers, and has stronger and more permanent positive impacts on its served area and often on an entire city. BRT has physical limitations: in cities with high density where exclusive bus ways or lanes cannot be provided, it cannot enter the urban core. LRT can do that either by using short tunnels or by operating on various types of ROW B, including running through pedestrian areas.

The major characteristics are listed and described in Table 4.14.

SUMMARY

This chapter is mainly based on the planning of public transport in urban areas. The first part deals with the urban transport planning based on the sustainable transport infrastructure of some cities. It also covers the classical transport models with different types and examples. This chapter also consists of information about travel demand estimation and its use. Another part shows some highlights on routing and scheduling of public. Also, it sheds some light on stakeholder consultation and its different methods. It also covers the high-order modes like BRT and metro rails and how they compare in three ROW categories.

TABLE 4.14
Characteristics of BRT and LRT

Characteristics/Mode	BRT	LRT	Superior Mode
Vehicle performance and passenger comfort	Good	Excellent	LRT
Investment cost	High	Very high	BRT
Implementation time	Short	Medium	BRT
Operating cost	Lower for low passenger volume	Lower for high passenger volume	-
System image and passenger attraction	Good	Excellent	LRT
Air pollution and noise	Considerable	No direct pollution, very low noise level	LRT
Interaction with land development	Limited	Very good	LRT

5 Planning and Management of Urban Transport

5.1 DEDICATED ORGANIZATION FOR CONTINUOUS URBAN TRANSPORT MANAGEMENT

Urban areas in India suffer from inadequate road capacity leading to congestion and delays in both the morning peak hour and evening peak hour. This peak period is getting extended for more than two to three hours in the morning and another two to three hours in the evening. This problem is much more critical in the metropolitan cities of India. The congestion is due to demand being much higher than the capacity of the road network that is available to cater to the demand.

Urban transport is an important infrastructure essential for urban growth and sustainability. In India, transportation demand in urban areas continues to increase rapidly as a result of both population growth and changes in travel patterns. Globalization has its impact on urban growth and its planning. The competition between Bengaluru, Chennai, Hyderabad, etc. to attract capital and the software industry is well known. At a global level also, this trend is evidenced and city governments (e.g., Sydney, Singapore, and London) have been adapting to their new global environments. It has been assumed that to gain a comparative advantage, it is necessary to beat competitors in the game of attracting investment from the leading sectors of the new globalized economy. Transport with its direct and indirect impact on environment, safety, and energy considerations has a vital role in the globalized city. To make a city competitive, livable, and attractive, a well-functioning urban transport system is essential. The contribution of a well-functioning urban transport system to the City Development Index (CDI) is around 30%. An efficient urban transportation system can increase society's productivity and welfare by improving people's access to jobs, recreational activities, and educational institutes.

5.1.1 Problems of Urban Transport and Consequences

Due to rapid urbanization, there is exponential growth of travel demand and doubling of trip lengths but only a marginal increase in mobility levels. There is no integration of land-use planning and transport. The increased urbanization and concentration of population in large cities are putting heavy pressure on the already over-saturated urban transport network, adversely affecting the productivity in urban areas. The mass transport systems are inadequate and are unable to cater to the demand imposed on them. As a result, there is a tremendous increase in the use of personalized vehicles, but the carrying capacity of roads has not kept pace.

DOI: 10.1201/ 9781003740247-5

This is leading to congestion, continuous slowing down of average vehicular speeds, increasing air and noise pollution, increasing accident rates, and excessive use of nonrenewable energy. Mass transit requirements of the metro cities and other large cities need to be addressed on a priority basis. With the growing traffic congestion, rush hour traffic is slowing to a crawl. This in turn leads to higher oil consumption and emissions, which are polluting urban areas beyond acceptable limits. There is an urgent need to reduce the total number of vehicles, reducing the total number of kilometers travelled, and reducing congestion, in order to give smooth flow of traffic to achieve better vehicular speeds, reduce pollution, and travel time. The problem of the pollution faced in Delhi and the measures initiated at the insistence of the Hon. Supreme Court in switching over to CNG as a fuel for buses are well known. Building on such initiatives, adopting cleaner technologies like electric vehicles (EVs) can further cut emissions and dependency on nonrenewable fuels, easing environmental pressures in crowded urban areas. A consequence of congestion is long travel times, which adversely affects the productivity of the vehicles and the productivity of the commuters. The road accidents have also been increasing causing loss of human lives.

The urban areas in the country confront a historic transportation crisis that has become a perennial war against increasing mobility, roadblocks, and air pollution. Given the financial constraints and environmental concerns, there is a need to adopt a comprehensive strategy for achieving sustainable urban transport. There is a need for a strategic land-use development policy in all large cities with an aim to integrate land use with mass passenger transport planning models to guide the future growth process of cities. Mass transport systems, both rail and road based, need to be developed. Considering the growing urbanization and its impact on overall conditions, intraurban and suburban rail transport, MRTS, and suburban rail systems are required to be integrated.

5.1.2 Efficient Use of Existing/Proposed Infrastructure

The emerging policy on urban transportation planning issues is the newly promulgated TSM action program. It is now required that TSM projects be included in the transportation improvement program (TIP) as the short-range element of the overall plan. Generally, in all cities with more than 2 lakhs of population, these TSM plans are compulsory while preparing comprehensive transportation plans.

The TSM considers a wide range of actions with low capital investment requirements that can improve transportation in the short term. It reflects the following:

1. Steeply rising costs.
2. Environmental concerns.
3. Intense competition for available resources.

It also makes it imperative that better and more efficient users for existing investments in the transportation infrastructure be found before additional investments are made in costly new facilities. The new concept is therefore to make more efficient use of the highways and transit systems already in place, thus reducing the need for new capital investments and for operation assistance.

The TSM on implementation provides the following advantages:

1. Fiscal economy
2. Better balance among the various elements of the urban transport system.
3. Broader local and national goals of energy conservation.
4. Environmental improvement.
5. Equity for transit dependents.
6. Urban preservation.

The TSM actions recognize the following features:

1. Characteristics of the existing transportation system.
2. Total demand for the transportation purpose in question.
3. Geographic distribution.
4. Nature of urban environment.
5. Locally determined goals.

For example, a pooled riding system in privately owned vehicles may serve the work trip demand most efficiently in a low demand density sector, while the conventional bus transit may serve in another sector. The high-occupancy commuter rail may prove efficient in another sector. The TSM plan fosters deliberately the use of a combination of modes that best represents an area's desired balance between the goals of efficient mobility, environmental quality, and social equity in the operation of a transportation system. The TSM plans are discussed covering the following aspects:

1. Warrants for implementation.
2. Prerequisites.
3. Advantages and disadvantages.
4. Implementation considerations.
5. Significant interrelationships among actions.

5.1.2.1 Broad Classification of Methods

All the TSM actions can be broadly classified into three groups:

1. Making more efficient use of existing infrastructure.
2. Reducing automobile usage in congested areas or time periods.
3. Improving transit service and transit management efficiency.

These three groups are generally dealt with in the following action program.

1. *Improved Vehicular Flow*
   a. Improvements in signalized intersections.
   b. Freeway ramp metering.
   c. One-way streets.
   d. Removal of on-street parking.
   e. Reversible lanes.

   f. Traffic channelization.
   g. Off-street loading.
   h. Transit stop relocation.

2. *Preferential Treatment of High-Occupancy Vehicles*
   a. Expressway bus and carpool lanes and access ramps.
   b. Bus and carpool lanes on city streets and urban arterials.
   c. Bus preemption of traffic signals.
   d. Toll policies.

3. *Reduced Peak-Period Travel*
   a. Work rescheduling.
   b. Congestion pricing.
   c. Peak period truck restrictions.

4. *Parking Management*
   a. Parking regulations.
   b. Park and ride facility.

5. *Promotion of Nonauto or High-Occupancy Auto Use*
   a. Ridesharing.
   b. Human powered travel modes.
   c. Auto restricted zones.

6. *Transit and Paratransit Service Improvements*
   a. Transit marketing.
   b. Security measures.
   c. Transit shelters.
   d. Transit terminals.
   e. Transit fare policies and fare collection techniques.
   f. Extension of transit with paratransit services.
   g. Integration of transportation services.

7. *Transit Management Efficiency Measures*
   a. Route evaluation.
   b. Vehicle communication and monitoring techniques.
   c. Maintenance policies.
   d. Evaluation of system performance.

In this chapter, it is proposed to discuss only those actions which relate to actions that improve vehicular flow and those that encourage HOVs.

5.1.2.2 Improved Vehicular Flow

The major goal of all TSM actions is to improve vehicular flow by implementing low-cost measures that increase the efficiency of existing road space and thereby avoid the need for roadway expansion. The following actions approach this goal by modifying and/or altering the designated use of road space.

1. Improvements in signalized intersections.
2. Freeway ramp metering.

3. One-way streets.
4. Removal of on-street parking.
5. Reversible lanes.
6. Traffic channelization.
7. Off-street loading.
8. Transit stop relocation.

If these actions serve to encourage increased use of low-occupancy autos, then the purpose is negated. Thus, only if these actions are integrated into a comprehensive TSM plan, will lasting benefits accrue.

5.1.2.2.1 Improvements in Signalized Intersections

Factors influencing the discharge of traffic at signalized intersections are signal cycle length and network control. Hence, by determining the most efficient use of signal time for each intersection, the amount of time available for individual movements at a group of intersections can be optimized and vehicular throughput can be increased. When, for example, 15 different cycle times were analyzed by SIGOP (Signal Optimization Program), a reduction in cycle was proved to improve significantly the average speed throughout. But if the cycle is shortened below 40 seconds, the pedestrian crossing time became unacceptable and hence was not desirable. In the case of pretimed and traffic actuated signals, research has shown the following facts:

1. Signal time phasing is not a controlling factor when narrow lanes, car parking, and right turns conflict with each other.
2. On major commuter routes through or adjacent to a Central Business District (CBD), coordinating timing of demand actuated signals can improve operational efficiency.
3. When signal time phasing is instituted in favor of a preferred route, the flow of commuter traffic is improved.
4. In small cities, signal maintenance is found to be inadequate. For effective performance, a city staff technician must be adequately trained, or the task assigned to an appropriate agency, or the task contracted to a fully qualified outside firm.

While going for traffic-adjusted computer control signals, the following stage-wise improvements are recommended:

Stage 1: Progressive connection of signals.
Stage 2: Manually changing the fixed time rate of pretimed signals.
Stage 3: Altering the demand responsive cycle of traffic actuated systems.

The most efficient control can generally be achieved through a computerized traffic responsive control system. In these processes, electrical impulses from traffic sensors that record traffic conditions at intersections in the network are fed into the decision program of a computer, which then selects and initiates the most efficient set of signal times and patterns for the network.

5.1.2.2.2 Expressway Ramp Metering

Signalizing expressway ramp intersections to control or meter the entry of vehicles into the expressway system is an increasingly popular and successful way of improving the use of existing facilities and improving the overall capacity. The following are the two different techniques adopted in practice.

1. Control volume of vehicles entering expressway so that expressway capacity is not exceeded and excess capacity is diverted to alternate parallel routes or less congested routes.
2. Smoothing of traffic flows as vehicles are allowed to enter only when they can be merged between platoons in the traffic streams.

The following equipment is usually employed for this purpose:

1. Gap detector.
2. Merge detector.
3. Input detector.
4. Queue detector.

The expressway ramp metering is applicable up to a lane capacity of 40–50 vehicles per km.

5.1.2.2.3 One-Way Streets

A one-way street system is one of the powerful techniques for increasing the capacity of streets as well as safety. The advantages of one-way streets are as follows:

1. Decreases the number of conflicts at intersections and, hence, improves safety and capacity.
2. More efficient use of capacity than two-way operations.
3. For narrow roads with a width of less than 15 meters without markings and with parking on both sides, maximum capacity can be achieved via one-way operation.

It is recommended that the following is the order in which this technique can be introduced. First remove on-street parking followed by the introduction of one-way streets and then proceed to signal preemption. Experience in other countries has shown that by introducing one-way streets:

1. Pedestrian accidents can be reduced by 20%.
2. Travel time can be reduced by 22%.
3. Vehicular stops can be reduced by 65%.
4. Cross-town delay can be reduced by 40%.
5. Signal time can be decreased from 50–90 to 35–60 seconds.
6. Can be integrated with contraflow and bus lanes.

The following disadvantages are, however, observed:

1. Increased walking to bus lanes.
2. Increased vehicle kilometers of travel.
3. Business reaction but improves the business in due course.
4. Pedestrian safety impeded due to increased speeds.

The following are the implementation considerations:

1. Signalized intersection performance can be improved by one-way streets.
2. Reversible lanes and contraflow lanes can be effectively employed by one-way streets.
3. One-way streets make effective utilization of traffic channelization.
4. Bus preemption at traffic signals can become easy.
5. Bus/carpool lanes can be introduced.

Figures 5.1–5.3 show how the number of conflict points at an intersection decreases by making streets one-way.

5.1.2.2.4 Removal of On-Street Parking

Removal of on-street parking helps to improve vehicular flow by increasing street capacity. By discouraging the use of automobiles in crowded CBDs to permit transit/paratransit usage, the institution of on-street parking regulation constitutes an important element of an area's overall parking policy. However, peak period parking restrictions are more difficult than enforcement of 24-hour restrictions and violations and interferes severely with peak period traffic.

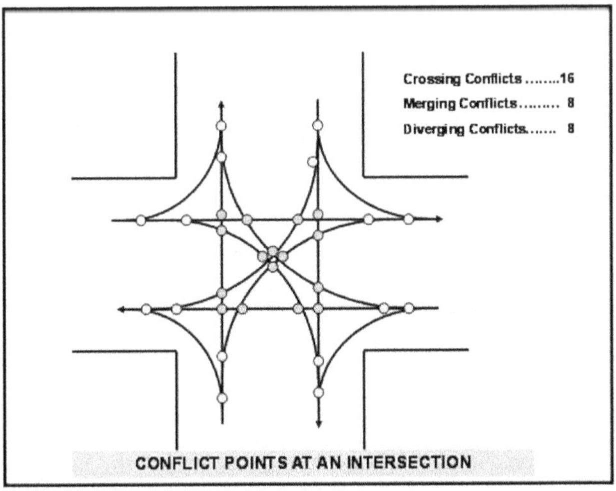

FIGURE 5.1 Conflict Points at an Intersection with Two-Way Movement

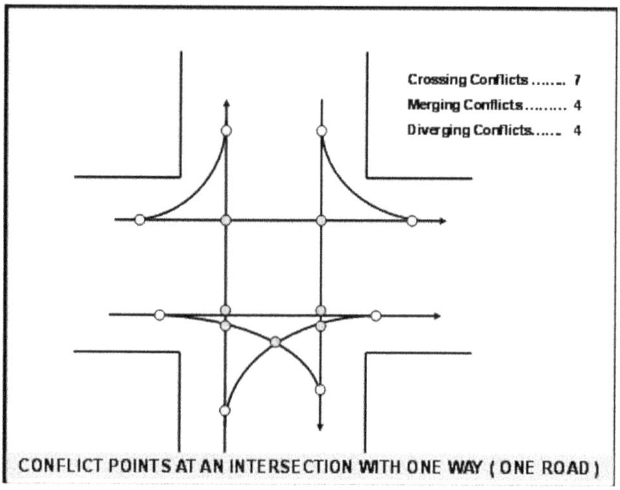

FIGURE 5.2 Conflict Points at an Intersection with One-Way (One Road)

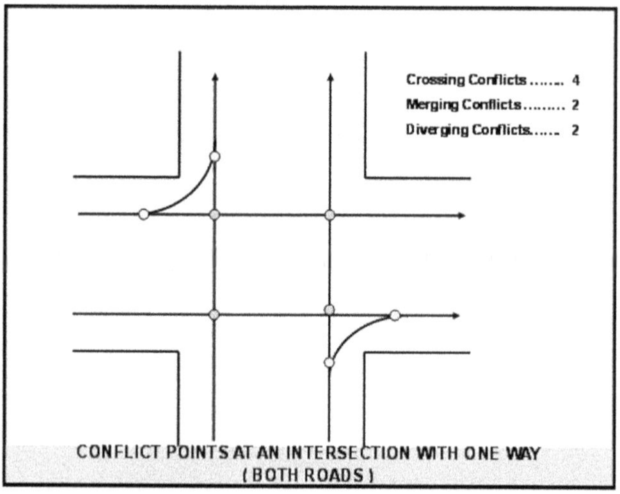

FIGURE 5.3 Conflict Points at an Intersection with One-Way (Both Roads)

The following are the advantages:

1. Substantial increase in roadway capacity is observed. Only 55%–65% of roadway capacity is utilized on two-way streets where on-street parking is permitted. When parking is eliminated, roadway capacity is approximately twice as great as if parking were permitted.

2. Improved traffic speeds of the order of 8 kmph are observed. Reduction in peak period travel to the time of 25% for auto, 10% for local buses, and 50% reduction in traffic delays are observed.
3. Removal of on-street parking also tends to greatly improve safety conditions, as it is a causal factor in 18–19% of all urban accidents.
4. Additional space created by parking banks improves efficiency of TSM actions. For example, the left/right turn lanes to relocate transit stops to midblock reduce hazards to cyclists. Bike lanes can also be introduced.

Strict enforcement is generally required to achieve maximum benefit from parking controls. Penalties for violating are to be severe for successful implementation. In most of the cities, 40–85% of all parking is on-street. In big cities, particularly in CBDs, only 15% is for parking. Hence, parking restrictions in small cities can improve efficiency substantially. Parking bans in CBDs have an adverse effect in CBD business in smaller cities.

5.1.2.2.5 Reversible Lanes

When directional flow is unbalanced by 65:35 split, reversible lanes can be used during peak periods to increase the capacity of the roadway in the directions of flow. They are applicable to roads not separated by a median strip. They are inexpensive and permit additional peak period directional capacity without adding roadways.

The experience says that peak period travel is reduced by 25%; violation of reversible lanes is a common problem compounded by lack of policing capability along the routes. Signing and lane-use control devices, overhead signal, and lane signals represent the basic costs for reversible lanes. Maintenance/enforcement costs are to be included as permanent costs in the estimates.

The advantages of reversible lanes are in improving the safety, reducing peak period congestion, and reduced travel time by 20%. The disadvantages are when only one lane is available for off-peak direction flow, long queues can form behind turning or disabled vehicles. On undivided roads, the following warrants may be useful for applying reversible lanes:

1. Sixty-five percent in one direction during peak flows.
2. The roadway capacity in the peak direction is adequate.
3. There is continuity in the route and in the width of the street.
4. There is no median, or left turns, and parking is restricted.

Progressive signal timing can improve the system with an interconnected signal controller for peak hour and for off-peak hour. Figures 5.4 and 5.5 depict the concept of reversible lanes.

5.1.2.2.6 Traffic Channelization

Channelization is a process of directing traffic into defined paths on roadways (Figure 5.6). It can reduce or eliminate potential hazards to motorists and pedestrians and avoid confusion in traffic flow. Generally, it requires islands, pavement markings, or other suitable means of facilitating traffic flow to separate potential areas of conflict.

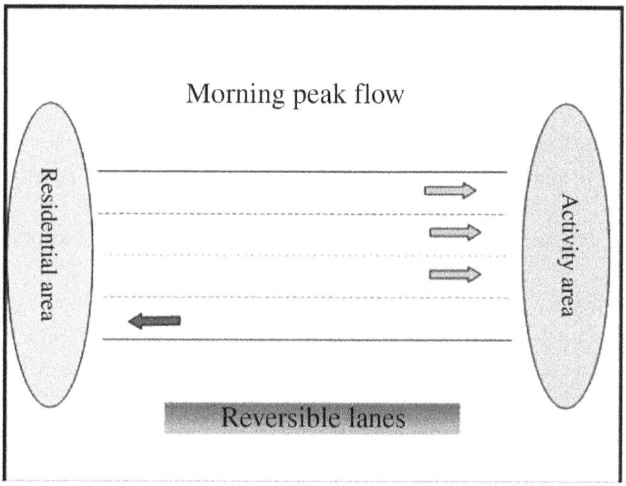

FIGURE 5.4 Reversible Lanes (Morning Peak)

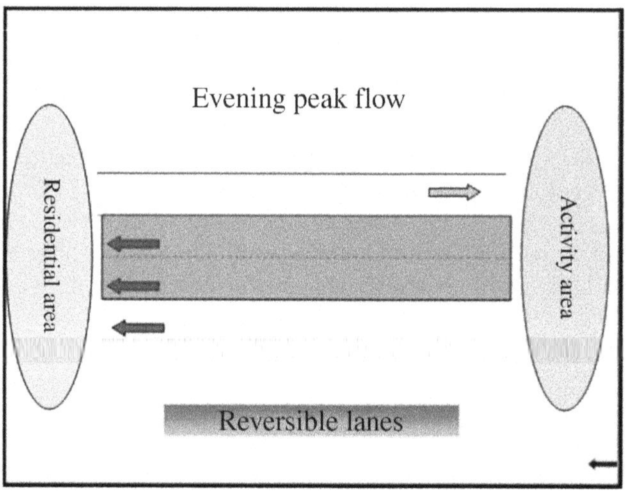

FIGURE 5.5 Reversible Lanes (Evening Peak)

If properly designed and executed, it improves safety and provides for maximum utilization of available space. It regulates movements and speeds of vehicles and can increase the speed up to 15 kmph. Where turning volumes are heavy, specially designated turning lanes can promote smooth vehicular flow by removing turning movements from through lanes. Right turn lanes, when accompanied by separate turn signal indicators, are highly effective in decreasing the right-turn vehicle exposure to oncoming through traffic. Left turn on red provision and special turning signals can increase the effectiveness of right turn lanes. Most of the cities have introduced

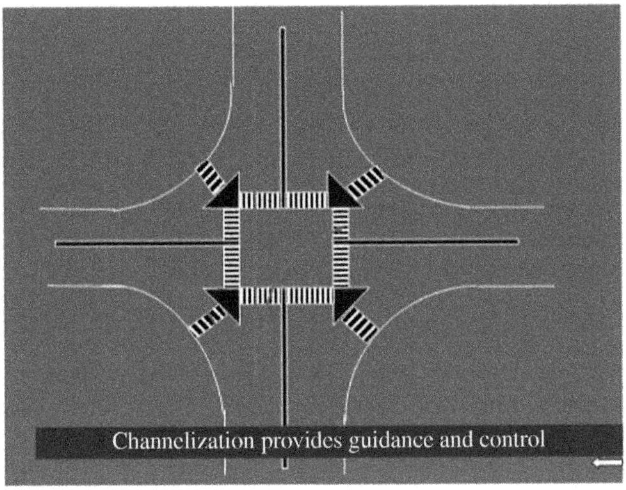

FIGURE 5.6 Channelization at Junctions

channelization projects. They are noncapital intensive, but costs vary between raised barrier and painted islands.

5.1.2.2.7 Turn Restrictions

When turning movements conflict with through traffic, turn restrictions may provide a suitable solution. Basically, turning bans on left minimizes vehicle pedestrian conflicts, while right turn minimizes vehicle-to-vehicle conflicts. Part-time turn restrictions can be effective during peak periods. However, public compliance is reportedly better where turning restrictions are maintained full time.

The following are the warrants for turn restrictions:

1. The right-turn volume exceeds 20% of total approach volume.
2. Right turn constitutes 10% of total movement on a given street.
3. Right-turn movements interfere with straight through movements of 1500 vph regardless of number of lanes.
4. A left/right movement interferes with pedestrian crosswalk volumes more than 2000 persons per hour.
5. Six-hundred vehicles conflict with 1000 or more pedestrians per hour.
6. Seven vehicles turn per green interval for several successive signal changes.
7. More than three intersection accidents that involve turning vehicles occur within a 12–month period.
8. The number of traffic lanes available at the intersection will accommodate only a single movement in each direction, and there is an appreciable demand for right turns.
9. Periodic gaps in opposing traffic do not occur due to traffic signals ahead.
10. Drivers will observe the right-turn prohibition when an alternative route is available.

5.1.2.2.8 Off Street Loading

Although preventing curbside truck loading and unloading removes traffic impedance, providing off-street loading facilities in CBDs is generally too costly to be included in the TSM plan. Thus, many cities can only modify urban zoning ordinances to require that new developments provide facilities for off-street loading and unloading and/or restrict or prohibit loading and unloading during peak hours.

5.1.2.2.9 Transit Stop Relocation

Curb side loading and unloading of transit and paratransit passengers, especially at intersections, can seriously impede traffic flow. Turnout bays near transit route junctions provide an ideal solution. However, in many situations, this solution may not be possible. In these cases, relocating transit stops may provide an equally effective solution. Basically, there are five places where transit stops can be located:

1. The near side of an intersection.
2. The far side across an intersection.
3. The far side of an intersection after a left turn.
4. Midblock.
5. Turnouts.

For the most part, bus stops are located at intersections. However, where blocks are exceptionally long or where bus patrons are destined for midblock employment or residence relocation, the transit stop relocation to midblock would be a suitable alternative, especially when accompanied by a midblock crosswalk. Due to visual obstructions created by the bus, motorists may find site conditions difficult at near-side stops. Also, it is hazardous for other vehicles when bypassing the loading bus. However, if the bus waits for a red light, nearside bus stops can reduce travel time. Far-side stops are advantageous where:

1. Other buses may turn in either direction.
2. Turn movements, especially left turns from the arterial in the direction being considered is heavy.
3. Cross traffic is heavy, and the curb lane and other lanes are needed for storage when signal is red.
4. Several streets meet at the intersection.
5. Site distance favors far-side locations.
6. Left turn-only lane effectiveness is increased with far-side locations.

The experience has shown that simple changes in bus stop locations have improved travel times to a great extent (Figures 5.7 and 5.8).

5.1.2.3 Preferential Treatment of High-Occupancy Vehicles

The following TSM actions improve the flow of high-occupancy vehicles during peak travel periods:

1. Bus and carpool lanes on city streets and urban arterials.
2. Bus preemption of traffic signals.
3. Toll policies.

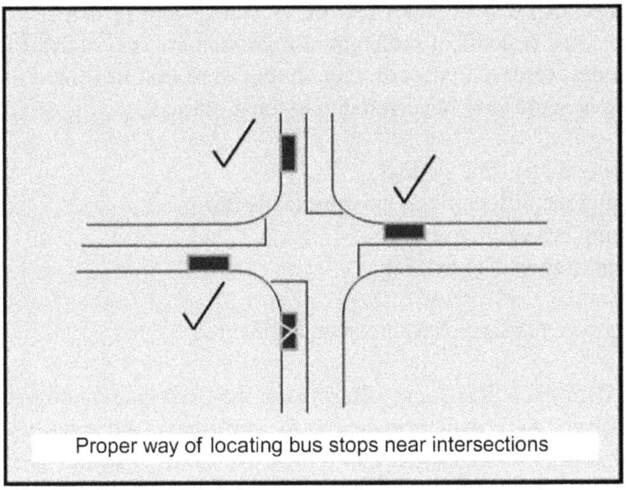

FIGURE 5.7 Bus Stop Relocation

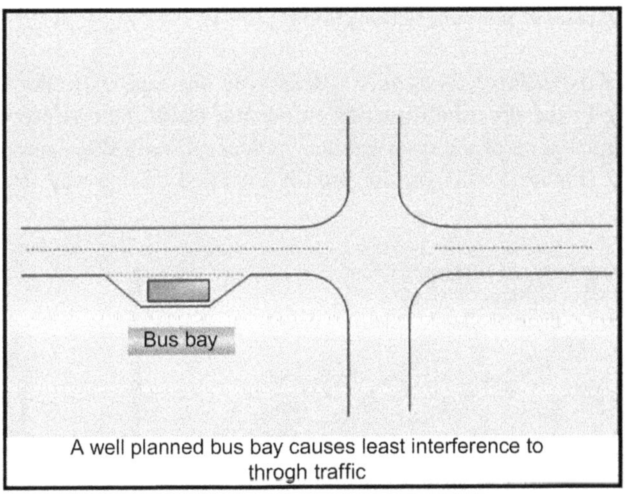

FIGURE 5.8 Bus Bays

The time savings and improved transit service levels resulting from these actions enhance the attractiveness of high-occupancy models. Ultimately, if these actions are incorporated into other TSM actions to encourage high-occupancy travel modes, significant shifts from low-occupancy vehicles and overall improvement in vehicular flow can be expected.

5.1.2.3.1 Bus and Carpool Lanes on City Streets

Reserved lanes on city streets and urban arterials can expedite the movement of high-occupancy vehicles through congested areas in or near the center of the city.

When dedicated expressway lanes can be extended onto principal urban arterials and city streets, the benefits of preferential treatment are optimized. Like dedicated expressway lanes, reserved lanes on city streets were first instituted for bus traffic. The following benefits have occurred due to these actions:

1. Improving travel time savings.
2. Increasing the utilization of existing facilities.
3. Improving bus service levels.
4. Reducing stop and go driving.

The following types of lanes have been experimented:

5.1.2.3.1.1 With Flow Bus Lane Reserving the curb lane on city streets for bus traffic is relatively easy and inexpensive to implement and requires a minimum number of changes in established bus routes. However, it makes it easy for other vehicles to violate the lane restriction and thus enforcement of the reserved lane concept is necessary to maintain its effectiveness. In addition, unless left-turn restrictions are instituted concurrently, conflicts with left-turn vehicles result. This entails removal of on-street parking and prevention of curbside vehicle stops during the time when reserve lane restrictions apply.

5.1.2.3.1.2 Contra Flow Bus Lanes Reserving one lane of a one-way street for buses moving in the opposite direction to normal traffic flow is proving to be an effective manifestation of the reserved lane concept. Contra flow lanes are virtually self-enforcing (Figure 5.9). They are usually installed on one-way streets carrying

FIGURE 5.9 Contra Flow Bus Lane

the light flow direction of peak period travel, which minimizes delays at signalized intersections during peak periods. However, when contra flow bus lanes are installed on one-way streets, a conflict with right-turning traffic, which the implementation of the one-way street had eliminated, is created. This type of contra flow lane will require the removal of on-street parking, and the prevention of curbside vehicle stops during affected hours.

5.1.2.3.1.3 Median Bus Lanes Median bus lanes are frequently instituted in the ROW, previously assigned to street car traffic (Figure 5.10). While this practice constitutes an efficient use of roadway space, all passengers must cross traffic to board or leave buses. An interference with right-turn traffic is also created.

5.1.2.3.1.4 Bus Streets Bus streets represent the full development of the urban roadway reserved bus lane concept (Figure 5.11). They are gaining acceptance in large cities. They provide relatively easy access to and egress from buses because buses are not competing for street space with autos. The bus streets are largely self-enforcing. The disadvantage is that auto and truck access to adjacent buildings is eliminated and, as such, alternate parallel traffic routes may be required.

5.1.2.3.2 Bus Preemption of Traffic Signals

About 10–20% of bus travel time on CBD routes is spent waiting at traffic signals. In order to reduce this delay, signal systems that allow the bus to preempt normal traffic signal timing patterns have been developed. Signals are equipped with optically or

FIGURE 5.10 Median Bus Lane in Ahmedabad (Janmarg)

FIGURE 5.11 Bus-Only Street in London

electronically actuated detectors that respond to signals transmitted from a device on the bus. This device causes the signal phase selector to lengthen the green interval or to shorten the red interval so that the bus can proceed uninterrupted through the intersection. Once the bus has passed, the signal returns to normal position. The following advantages were noticed when installed:

1. Reduction in bus travel time.
2. Increased auto delay.
3. Impediment to auto traffic flow.
4. Platoon preempting is better than four to five buses cross.
5. Ten to 15 buses per hour are warranted for installations.
6. This is more effective with contra flow.

5.1.2.3.3 Toll Policies

Preferential treatment at toll collection points can be accorded to high-occupancy vehicles by permitting nonstop passage at tolls.

1. By differential tolls favoring high-occupancy vehicles.
2. Preferential treatment at toll collection points can be accorded to high-occupancy vehicles by permitting nonstop passage at tolls by displays.

5.1.2.4 Reduced Peak Period Travel

The following TSM actions discussed can improve peak-period flow by reducing the number of VKT during peak periods:

1. Work rescheduling.
2. Congestion pricing.
3. Peak period truck restrictions.

Altering work schedules would eliminate the need for some workers to travel during peak periods. Imposing higher charges during peak periods, for example, tolls, transit fares, and roadway fees, would make it more costly and hence less desirable to travel at peak times. Finally, restricting the times when trucks can pick up and deliver merchandise to off-peak periods would remove impedance to peak period traffic flow. Other complementary TSM actions could be:

1. Increased work trip occupancy through preferential treatment of high-occupancy vehicles, ridesharing, important services, etc.
2. Instituting higher parking fees for long-term parking (congestion, pricing).
3. Encouraging walking and bicycling for short work trips.

5.1.2.4.1 Work Rescheduling

The rescheduling of work hours to reduce the amount of work-related peak period travel could be accomplished by the following:

1. Staggering or rearranging employee starting and quitting times to achieve a more even distribution of arrivals and departures over a longer period of time.
2. Reducing the work week to four days, and as is commonly practiced, increasing the number of hours worked per day, staggered or flexible work hour programs.

In the past, work reschedules were implemented primarily to reduce peak period congestion at company sites or on transit facilities. The cost of implementation is negligible. However, some costs for data collection and processing of surveys that are included in evaluation procedure can be implemented. The advantages are as follows:

1. Readily implemented at relatively low cost.
2. Can be implemented on an area-wide basis.
3. Peak hour reduction—the need to construct new roads will be delayed.
4. At other places like restaurants, lobbies, elevators, and stores, queues can be reduced.
5. Employees experience shorter work trip travel times and more planned and better work conditions.
6. Flextime experience improves an employee's morale as the employee is given some control over his/her work schedule.
7. An indirect benefit on energy consumption and/or air quality is observed.

Implementation considerations:

1. No difficulty in rescheduling work by 15 minutes by all companies.
2. Thirty minutes per hour is reliable only by large firms or public agencies.
3. Riding pools may be affected.
4. The actions must be coordinated with local transit authority; transit operators can help identify the locations themselves.

5. If traffic reduces drastically, then it may encourage single-occupancy auto-counter production.
6. Employees' proximity to each other can be effective.
7. Predicting the effect on roads is complex due to trip length—this may increase congestion also.
8. Instituting regional citywide staggered work is found to be counterproductive.

5.1.2.4.2 Peak Period Truck Restrictions

Congestion charging is used as a measure to disincentivize use of private vehicle and encourage public transport and NMT in busy areas of a city, particularly during peak periods. This is done by cordoning the city centers and charging all private vehicles that are entering this area. While implementing congestion charging, it is important to provide good alternatives for private vehicles for this measure to be successful. This is another TSM action that can be implemented by the following:

1. Instituting truck roads.
2. Preventing truck travel on same routes.
3. Establishing temporal restrictions to accommodate merchants' desire of 7 AM to 6.30 PM.
4. Restricting all truck pickups and deliveries to nighttime.
5. Permitting trucks only during off-peak periods.
6. Restrictions during peak periods are also a problem.

5.1.2.5 Promotion of High-Occupancy and Non-Vehicular Travel Modes

This TSM action is implemented through the following:

1. Ridesharing.
2. Encouraging human-powered travel modes.
3. Auto restricted zones.

They improve the efficiency of existing transportation by reducing the number of vehicles—kilometers traveled in congested areas or time periods without altering mobility.

5.1.2.5.1 Ridesharing

It is a form of paratransit that entails prearranging of shared rides for people traveling at similar times from approximately the same origin to approximately the same destination. In lightly populated areas, transit efficiency is not profitable and resorting to ridesharing is found to be a feasible TSM action. The following are the variations:

1. *Ridesharing:* Car owners drive others by rotation
2. *Carpools:* Organized system. No parking fee is charged in these cases (Figure 5.12).
3. *Van Pools:* Over 10 miles distance, one-way trip distances are much cheaper than carpool system when 10–12 persons are carried.

FIGURE 5.12 Promotion of Carpooling

Area-wide pooling programs can also be considered by discussion with public transport companies, and the best-suited program can be implemented.

5.1.2.5.2 Human-Powered Modes

Encouraging bicycles, etc. is found to have the following benefits:

1. Health.
2. Recreational.
3. Congestion reduction.
4. Prevention of air pollution.

To encourage bicycles bikeways can be constructed (Figure 5.13). They are classified as:

Class I: Completely separated ROW designated for exclusive use.
Class II: Restricted ROW exclusive use or semi exclusive use of bicycles. By signs and street markings vehicle parking may be allowed.
Class III: A shared ROW designated by signs placed on vertical posts and stenciled on the pavement.

Usually, Class III is only feasible. In all these cases:

1. Convenient secure storage of parking is important.
2. Parking to be located near the major trip generator.
3. Change of mode terminals.
4. Bicycle lockers, racks at transit stops (Figure 5.14).
5. Transit/bicycle piggyback can be tried. Bike bus carries—second trailer carrying cycle onboard.

FIGURE 5.13 Bicycle Lane in Ahmadabad

FIGURE 5.14 Secured Bicycle Parking Facility Near Bus Stop

The transit vehicles, condition of street surface lighting, conflicts with MV, etc., are the major issues that are to be seriously managed.

Improving the existing pedestrian circulation system can be achieved by the following actions:

1. Widening sidewalks.
2. Clearing of impediments to pedestrians.
3. Better lighting.
4. Midblock crossing arrangements.
5. Access to transit/parking garages.
6. Arcades.
7. Malls.
8. Plazas.
9. People mover system.
10. Only pedestrian streets for people by providing plants, fountains, kiosks, etc.

Generally, it is recommended to provide 1.2 m width for paths for residential streets and at least 1.8 m for commercial areas. One can add +0.6 m for obstructions on pedestrian paths.

5.1.2.6 Parking Management

No other operational control can have as dramatic an effect on traffic flow as the proper management of an area by parking supply controls. Decisions are important on the following:

1. Location.
2. The amount of on- and off-street space allocated to parking.
3. The parking charges apply to the allocated space.
4. The length of parking time permitted affects nearly all the TSM actions.

5.1.2.6.1 Parking Regulations

These are meant:

1. To control the number and type of vehicles entering congested activity centers.
2. To couple with other TSM actions to improve transit and paratransit service, such as preferential treatment for high-occupancy vehicles. The number of vehicles entering an area can be reduced.
3. To avoid the need for additional regulations by enforcing existing parking restrictions (as in many cities) more stringently before instituting new regulations.

5.1.2.6.2 Other Points

1. Long- and short-term parking—To discourage long-term parking by commuters and to favor short-term parkers.
2. Parking taxes—Imposing a citywide tax. In Los Angeles, for instance, a 2-year 25% tax on all parking was imposed.
3. Parking surcharges—Surcharge on certain users of parking facilities is another way. Employers can levy parking surcharges on employees with single passenger vehicles.

4. Increased parking rates—General increase rates can also control parking.
5. Interrelated effects—Mode shift is the after effect.

5.1.2.6.3 Implementation Considerations

1. Should be implemented fully over a larger area, otherwise spillovers take place.
2. These actions should encourage through-traffic.
3. Other actions like CBD transit circulation, etc. are to be examined before implementation.
4. Parking charges do not reduce parking as only a shift takes place.

5.1.2.6.4 Limiting Parking Space

While increased parking charges represent an economic disincentive, limiting parking spaces restrict supply.

1. Not an encouraging experience for bigger cities.
2. By restriction of parking spaces up to 45–51% of the demand by zoning laws can discourage parking.
3. Effective in small urban areas.
4. Adequate alternatives are to be considered.
5. Extend to 1/3 mile beyond the parameter of the activity center.

5.1.2.6.5 Park-and-Ride Facilities

Coupling fringe parking with an express transit service can bring good effects (Figure 5.15).

The space requirements for self-parks are about 350 sq. ft./vehicle.

Advantages:

1. Parking shifts to fringes.
2. Attractive/cheaper.

FIGURE 5.15 Commuter Park-and-Ride

5.1.2.6.6 *Guidelines*

1. Parking area to be large enough for circulation and for pedestrian safety.
2. Close to activity center (1 mile).
3. Located on dense travel corridor, radial, etc.
4. Not to compete with other service areas.
5. Adjoining areas not to be affected.
6. Convenient access/egress to buses.
7. Quick transit service, say 5 minute headway.
8. Free parking (if possible).

5.1.2.7 Transit and Paratransit Service Improvements

A number of aspects can be improved to enhance the LOS of transit/paratransit services through the following:

1. Transit marketing.
2. Security measures.
3. Transit shelters.
4. Transit terminals.
5. Transit fare policy and fare collection techniques.

In addition, the following facilities can help:

1. Extending the service provided by transit with paratransit operations on to a unified network of services.

5.1.2.7.1 *Transit Terminals*

These are the places where collection, distribution, and transfer functions are combined.

There are a variety of designs, but basically they can be classified as:

1. Park-and-ride is designed for transfer function and is suitable for small and low-density areas.
2. Mode transfer terminals are suitable for outlying areas.
3. Off-street terminals are useful for urban settings.
4. Central area transit terminals are interchange facilities.

5.1.2.7.2 *Central Area Transit Terminals*

Off-street passenger distribution is freed, and curb lanes are liberated. Multiuse terminals can encourage buses and transit systems.

1. *Outlying Terminal:* Aids in consolidating routes. Reduces vehicular traffic destined to city.
2. *Off-Street Terminal:* Downtown curb loading capacity is limited. On-street loading/unloading is slow.

5.2 DATA COLLECTION, RETRIEVAL, AND ANALYSIS FOR EFFECTIVE MANAGEMENT

5.2.1 Information Files and Data Collection: Survey and Counts

Effective planning and management of transit operations, scheduling, and analysis of efficiency in general require accurate data about operations and usage of transit lines and networks. Transit agencies must therefore organize, maintain, and regularly update systematic information and data. These data files should include the following:

1. Data about lines and facilities, such as locations and equipment of bus stops on streets, designs and dimensions of rail transit ROW, track layouts, signals, stations, and yards and workshops.
2. Data about public transport vehicles, including body dimensions, design and performance data, age, condition, and breakdowns.
3. Types of service provided and schedules for all modes and lines.
4. Data about usage of services, such as passenger boarding, alighting and volumes along the lines, their time fluctuations and their trip lengths, trip generation of different major activity areas, such as shopping malls, stadiums, university campuses, medical complexes, and others.
5. Miscellaneous information on crime incidents, methods of fare collection and types of fares, train-dispatching efficiency, passenger preferences with respect to schedules, riding comfort, and other service parameters.

5.2.2 Organization of Surveys

Good planning and continuity in data collection are necessary to obtain and maintain the current database. The frequency and comprehensiveness of field surveys must be determined based on a compromise between the need for accurate information and the cost of surveys. A good practice is to organize major, detailed surveys at longer intervals and supplement them by minor ones, often on a sampling basis, within these intervals. In other words, a comprehensive cross-sectional set of data, describing entire operations at one time, is combined with longitudinal or time data on key elements, such as passenger volumes on maximum load sections or peak-hour flows into the CBD.

5.2.3 Transit Speed and Delay Surveys

The purpose of a speed-and-delay survey on street transit lines is to find the distribution of time a TU spends in travel, classified as running, dwells at passenger stops, and several categories of delays. It is also intended to record the locations, durations, and causes of the delays. The measured travel times and delays are used to compute operating speed and reliability of service during different periods of the day. This information is important for planning possible operational improvements.

5.2.4 Passenger Volume and Load Count

The purpose of passenger counts is to determine the passenger volume on TUs over different sections of a line, the maximum TU load and the section on which that occurs, variations of volumes in time, analysis of service quality, that is, all elements needed for scheduling of operations. A detailed survey should include TU passenger load counts at several points along each line, particularly focusing on the sections with heavy loads to find the maximum load sections (MLSs). Minor annual surveys may be limited to MLSs and one or more additional sections on each line to verify the changes recorded on the MLS.

5.2.5 Passenger Boarding and Alighting Counts

The most detailed information on passenger volumes on a transit line is obtained through the counts of boarding and alighting passengers at each stop or station along its entire length. This count provides data on the number of passengers using each station, as well as TU loads at all points along the line. One can also compute the distribution of passenger trip lengths and total work of the line in passenger-km for any hour or day. This represents virtually all the information, extending or shortening of the line, addition or abandonment of some stops, etc.

5.2.6 Other Types of Surveys

In addition to the above-described rather standard surveys, many other surveys can be undertaken to obtain various types of data. Several examples of such surveys are described here.

1. Transfer counts.
2. Fare usage.
3. Passenger travel information.
4. Attitudinal and modal split survey.
5. Use of timetables.

5.3 ROLE OF PRIVATE SECTOR IN PUBLIC TRANSPORT MANAGEMENT

5.3.1 Experience of Private Operators in India

Generally, the service level provided by private operators is far from satisfactory, and there are frequent complaints of over-charging, rash driving, rude behavior of crew, nonadherence to scheduled timetables, not starting services until the vehicle was fully loaded, and very poor condition of buses. It is suggested that privatization must be accompanied by appropriate regulatory frameworks and institutions. Table 5.1 shows the percent of public and private sector buses in India.

TABLE 5.1
Percent of public sector buses in different years

Year	Public Sector Buses (in thousands)	Private Buses (in thousands)	% of Public Sector Buses	% of Private Buses
1976	52.2	62.8	45.4	54.6
1981	69.6	92.3	43	57
1986	84	143.3	37	63
1991	106.1	225	32	68
1996	111.1	338.7	24.7	75.3
2001	115	518.9	18.1	81.9
2002	114.7	520.3	18.1	81.9
2003	114.9	605.9	15.9	84.1
2004	111.4	656.2	14.5	85.5
2005	113.3	779.4	12.7	87.3
2006	112.1	879.9	11.3	88.7
2007	107.8	1242.5	8	92
2008	113.6	1313.6	8	92
2009	117.6	1368	7.9	92.1
2010	118.8	1408.3	7.8	92.2
2011	130.6	1473.2	8.1	91.9
2012	131.8	1544.7	7.9	92.1
2013	137.9	1676.1	7.6	92.4
2014	140.2	1746.7	7.4	92.6
2015	140.5	1830.3	7.1	92.9
2016	142.9	1613.8	8.1	91.9
2017	149.1	1715.1	[illegible]	92
2018	151.9	1790.6	7.8	92.2
2019	152.4	1896.5	7.4	92.6

5.3.2 Competitive Advantages of STUs

In the present situation, it is obvious that the development of well-coordinated public transport services requires both the STUs and the private sector bus operators to play a major role in the coming years. However, so far, the country has not been able to develop any major players in the private sector to run coordinated bus transport services. This sector is still dominated by small ownership fleets with very little investment in terms of infrastructure needed for running reliable and responsible public transport services. The levels of training and the working conditions of labor in the unorganized private transport sector are extremely unsatisfactory. Finally, the road safety record of small fleet operators with ill-trained workers cannot be said to be satisfactory. In such an environment, it is the STUs which give public transport an element of reliability and coordination. The STUs have much strength which needs to be built upon in the coming years. They have a vast pool of professional managerial talent with long experience in running public transport systems. They have also built up in most states extensive infrastructure facilities for bus operation on a large scale. The STUs have successfully operated uneconomical routes in rural areas cross subsidizing them with surpluses from the more remunerative routes. Indeed, there are examples of STUs in our country running some of the largest bus operation networks in the world. Despite operating buses manufactured with outdated technology, STUs have achieved very high rates of vehicle productivity, fuel economy, and manpower productivity. They have succeeded in training and employing a huge work force of skilled manpower which is approaching the one million mark in terms of directly employed workers. They have also achieved one of the main objectives of nationalization, namely ending exploitation of workers in the transport sector by providing terms of employment which conform to accepted norms of labor welfare and labor laws. The STUs are also playing an extremely valuable role in many states in promoting social and economic development. To cite one example, the spread of education in rural areas (particularly among girl children) is greatly aided by the concessional travel facilities offered to students by the STUs. It is our view, therefore, that these strengths of the public undertakings in this sector should be profitably used for developing public transport services in the country in combination with the private sector bus operators.

5.3.3 Private Transport Operators' Role in Selected States

5.3.3.1 General

Private operators have been playing a complementary role, and they have been aggressively occupying the space vacated by state operators. They have developed a wide network of operations and run through all major centers of the country. It has led to a generation of large entrepreneurial groups, like the road cargo sector. If we also include the unauthorized operators like jeeps, maxi-cabs, and even trucks, the reach of private operators is even more. They run even in interior rural areas and hilly areas. The major cause of this development has been the absence of organized large-scale players in the field. This field has been mostly occupied by unorganized small fleet operators barring some whom we can call at the most as medium-sized operators. The challenge to policymakers lies in reigning in these unorganized operators and making the sector attractive to large-scale players. We present below the industrial structure of private operators and their operational characteristics.

5.3.3.2 Rajasthan

In 1979, the Rajasthan government issued 64 permits to private operators to operate bus services alongside RSRTC buses in Jaipur. In two years, the private operator's share of buses increased to 122, which resulted in unhealthy competition forcing RSRTC to withdraw its services in 1984. Subsequently, to meet the increasing traffic demand, the state government issued permits to private operators to operate different kinds of services and, in 1990, Jaipur had 2750 auto-rickshaws, 223 Vikram taxis, 700 tempos, and 289 minibuses.

The service level provided by private operators was far from satisfactory, and there were frequent complaints of over-charging, rash driving, rude behavior of crew, nonadherence to scheduled timetables, not starting services until the vehicle was fully loaded, and very poor condition of vehicles and the resultant breakdowns and overloading were almost twice the capacity of the vehicle, etc. This situation compelled the Rajasthan state government to convene a meeting and directed RSRTC to restart its services in Jaipur.

5.3.3.3 Orissa

During the later part of the 1970s, private operators owned and operated 2160 buses in Orissa state, while the state-owned Orissa State Road Transport Corporation (OSRTC) operated only 712 buses. Between the period 1978 and 1988, the fleet strength of the private operators doubled while of RTCs fleet declined. OSRTC's fleet utilization was one of the lowest in the country during 1991–92. With the major share of buses with the private operators, they were very powerful and influential. While the private operators cornered 91.3% of the total profitable routes in the state, OSRTC was forced to operate all other nonremunerative routes spread all over the state. Taking advantage of this situation, the private bus operators demanded a fare increase from 10.0 to 20.0 paise per passenger–km, which the government could not agree to, forcing the private operators to go on strike.

5.3.3.4 Madhya Pradesh

In Bhopal city, largely private operators provided public transportation while the Madhya Pradesh State Road Transport Corporation (MPSRTC) was providing services only to government servants and schoolchildren at concessional rates. About 400 minibuses, 417 tempos, and 4600 auto-rickshaws were involved in providing public transport requirements in Bhopal city. As the fare structure in respect of various modes differed, public interest was the first victim in the ensuing heavy cutthroat competition. Major causes for the large number of accidents by private vehicles were rash driving, dangerous overtaking, and the poor quality of the drivers employed by the private operators.

5.3.3.5 Andhra Pradesh (Combined)

Private operators until 1978 provided the road passenger services in Vishakhapatnam. They were operating 44 routes with 124 buses. The bus operation was found to be tardy, and there were frequent accidents involving private buses. During the year 1978, the Visakhapatnam city witnessed a large-scale public agitation and law and order problems on account of the poor bus service in the city. The agitation compelled the state government to issue directions to the Andhra Pradesh State Road Transport Corporation (APSRTC) to introduce bus services in Visakhapatnam. Ultimately, APSRTC had to introduce 100 buses with effect from December 11, 1978, which was subsequently increased to 129 by December 20, 1978.

In Vijayawada, private operators provided passenger services. The traveling public was not satisfied with the services provided by the private operators with the result Vijayawada witnessed a series of agitations leading to law-and-order problems. Thus, APSRTC introduced bus services in Vijayawada.

Prior to 1987, private operators were providing bus services in the district of Srikakulam. There were numerous public complaints about private operators that their services were not reliable, not safe, not punctual, inadequate, and of poor quality, and that they were only interested in making quick profits, etc. This situation forced the Andhra Pradesh government to nationalize the bus services in Srikakulam in 1987. After the nationalization, the fleet of buses in Srikakulam started increasing on an average of 15.1% compared to the 5% increase before nationalization. With appropriate scheduling and an increased fleet, APSRTC operated 1.08 lakh passenger-km daily compared to 53,311 passenger-km per day operated by private operators prior to nationalization. A random survey revealed that the public in Srikakulam district are happy with nationalized passenger services of APSRTC compared to private operations.

5.3.3.6 Kerala

In Kerala, both the state-owned Kerala State Road Transport Corporation (KSRTC) and private operators provide passenger bus services side by side. While the private operators share of the total passenger buses in the state is around 89%, the KSRTCs share is only about 11%. Because of their major share in the total fleet of buses, the private operators are very much organized and are very influential. In the beginning of the 1990s, the private operators in the state went on an indefinite strike pressing for an increase in the existing fare structure. The strike went on for a couple of days, forcing the state government to increase the fare from 12.5 to 14.0 paise per passenger-km.

5.3.4 Factors Influencing Private Operators

The following factors in general influence most of the operators. These factors were identified based on discussions with a number of private operators located in Hyderabad, Bengaluru, and Pune.

5.3.4.1 Easy Entry/Easy Exit

The passenger road transport sector is characterized by free play for small-time operators, if we leave out the state transport services. Anybody, even with one bus, can start a service. They can almost operate at well. There are certain routes which have been denationalized in which they can operate. But they may just take license for one route and operate on some route which they find profitable. They start as contract carriers but operate more as stage carriers. They are also fleet footed. The moment they find a particular route unprofitable, they shift to another route. They can cut frequency and change routes depending upon traffic flow. In fact, even the larger players in this industry complain about these small fleet operators. They are also threatened by substitutes. Just as state transport operators complain about private operators, these unauthorized bus operators complain about maxi cabs and autos that are also unauthorized.

5.3.4.2 Low Investment/Low Technology

It is a low-investment industry, and bankers are willing to fund bus service operations. The performance of these buses should be good and that is why the bankers are willing to lend. The performance is good because these operators are not bound by any route restrictions or service obligations and pay lower taxes. Technology has been mostly stable except in recent times when Volvos started flooding the market. Maintenance is simple and can be outsourced. Often, they do not even maintain a garage.

5.3.4.3 Faster Pay Back

Unlike other services, bus services start fetching returns from day one. It generates cash daily and if we consider passes, these are in fact advances. That is why debt servicing is also not a problem for them. Their operations can in fact be characterized as highly liquid yet unviable. The main issue about these operators is that they lack any long-term stake in the sector. They are there to reap short-term gains. This attitude is antithetical to the management of public utilities.

5.3.4.4 Control

The dominance of small fleet operators has made any regulatory efforts a big failure. It is very difficult to monitor and track these operators. There is a complete lack of information about their operations. That is why it is difficult to monitor their schedules, frequencies, and traffic patterns. Most of these operators are proprietary concerns, and they maintain little physical or financial data. This nullified practically any critical analysis of their revenue model or their cost structure from our study point of view.

5.3.4.5 Revenue Model

It emerges from a study that organized operators mostly prefer interstate operations as that is the most profitable. Very few organized players are in the intrastate segment. They are very adept in identifying profitable routes and are flexible in scheduling. They are also aggressive in their pricing. They follow different pricing depending upon the peak hours and peak days. They also work through a network of booking agents. There are some states like Karnataka which have denationalized intradistrict routes. Here they are expected to operate on scheduled routes and frequencies.

It also emerges that these operators also operate on a thin margin. In Karnataka, it was found that the moment KSRTC allowed flexi-fare to its divisional controllers, the private operators could not face stiff competition from them. Similarly, they again found that the moment KSRTC introduced new buses, the private operators found it difficult to compete with them. Many of the operators simply moved away to other routes. They also could not cope with the recent fuel hikes, as the RTCs did not follow up with fare hikes. It also emerges that in each state there are very few operators with more than 50 buses. They all mostly own five to ten buses. This may be because recent hikes in fuel costs and RTCs stubbornness in not hiking fares has made many of these services unviable even for private operators. This again establishes the point that this sector is still unattractive for organized players. Even in cargo movement, we find organized players like TCI and TVS, but in bus services, we hardly find any organized players. They try to complement their earnings through revenues from unaccompanied luggage. This is especially lucrative in the case of interstate movement. The contribution from this source can range from 15% to 25% depending upon the aggressiveness of the operators. They work on the principle of earnings per bus and if it covers the cost of running, they are satisfied with it. Their earnings per kilometer may work out like this. Their average VU per month is about 600–625 km per day. They also assume revenue from luggage to the tune of 15–25% of the traffic revenue. Many operators opine that the latest high technology buses may not be viable on many routes. The business model of these high-tech buses is yet to be established. These may succeed as long-distance interstate operations or as tourist buses.

5.3.4.6 Cost Structure

The study sought to construct the cost structure of private operators but found it extremely difficult. It was observed that the operators hardly maintain any physical or financial information. Also, as these are proprietorships, they do not have to disclose financial information. They work with basic information like average vehicle utilization in a month, earnings per bus per trip, earnings through luggage, crew salaries, fuel efficiency, and bus cost.

We tried to normatively construct their cost structure. They provide for double crews on long distances. They are engaged beyond normal hours. They may not have conductors on these trips. Their fuel efficiency is low as they are not so focused on this as their maintenance is not professionally managed like RTCs. Excepting the large operators, rarely do they own any maintenance depots. Their overheads are low as they are mostly owner managed. They do not maintain huge office infrastructure. They maintain meager infrastructure for maintenance. They are at an advantage when it comes to MV tax as they pay per seat per quarter.

The cost per kilometer of Volvo and luxury bus has been worked out as presented in Table 5.2. We constructed this cost structure normatively based on the information supplied by some operators. On average, the cost structure is expected to behave on the pattern depicted in the table. It assumes that Volvo will be running 500 km per day and luxury bus 350 km per day.

Private Operators try to manage cost by controlling personnel cost, maintenance, and overheads. Their buses and depreciation costs have been going up as competition

TABLE 5.2
Economics of private operators

Component	Volvo Rs./km	Share of cost %	Luxury Rs./km	Share of cost %
Fuel	12.60	44.06	8.00	46.51
Personnel	2.10	7.34	2.50	14.53
Maintenance	2.80	9.79	1.75	10.17
Depreciation	5.60	19.58	1.85	10.76
Interest Cost	3.40	11.89	1.50	8.72
Misc	2.10	7.34	1.60	9.30
Total	28.60	100.00	17.20	100.00
Earning per km	36.00		21.00	
Profit per km	7.40		3.80	

is getting severe. They are competing on bus brands like Volvo, and interiors and facilities. The cost structure of organized operators is slightly different as they maintain a marketing office and have better maintenance infrastructure. Of course, they are not under any obligation to build any infrastructure for passenger amenities or bus stands or depots.

5.3.4.7 Quality Parameters

The stress on bringing in organized players is to usher in professionalism and concerns for quality. It may be fashionable to say that the market will determine the quality aspects of service. The absence of a credible alternative public mode has left the public with no option except to go for any service that is in operation. For example, they may not even have a spare bus in case of breakdown or accidents. They fail miserably on quality parameters like the following:

1. Bus conditions.
2. Reliability of service.
3. Crew behavior.
4. Accident rates.
5. Passenger amenities.
6. Maintenance infrastructure.

The services provided by private operators are a far cry, excepting select leading operators in each state as pointed out by the survey conducted by the Indian Institute of Management Bangalore (IIM) Bangalore (IIMB).

5.3.4.8 Competitive Edge

The competitive advantage of private operators is very clear. They are entrepreneurial and market driven. These operators are able to adopt this because as of now they are not burdened by any social service obligations. They are cost effective. Because these are entrepreneurial, they are cost conscious, and try to control buildup of overhead. The share of personnel cost is quite low compared to STUs. They may lose some of these advantages if they are brought under regulation and are asked to follow labor laws, quality parameters, and service obligations.

5.3.5 Policy Issues Confronting Private Operators

Discussions with private operators indicate that they are equally concerned about policy dilemmas. The large players among them feel that they are also providing a useful service, which goes unrecognized. They would like to have clarity on their role and what is going to be the policy on the privatization of routes. They feel piecemeal privatization will not help. In the absence of clarity on this, they are not definitely willing to commit resources to this sector. They are open to operating on intradistrict routes and intrastate routes if they are provided with appropriate policy and tax support. This is discussed under USO under policy section. They also would like to see a strong regulatory body, as within their group, organized operators face difficulties with unorganized operators. They feel they should control the number of players who can operate in each segment. Too many players can spoil the case. They are wary of concessional segment though they are willing to cater to them as long as the subsidy is compensated. They agree that they cannot provide infrastructure the way STUs have built up. They also opine that they should be given access to bus stands for which they can be charged a fee. They quote the Chennai model as the case.

5.3.6 A Relook at Privatization Options

The primary objectives of privatization in this sector will be to encourage competition and to encourage large-scale organized players to enter the field. This can only ensure the economy of operations, fresh injection of funds in the sector, and professionalism in the field. Otherwise, it will be RTCs at one end of the spectrum and a host of small fleet operators at the other end. Both groups are difficult to manage through government control mechanisms. It is important to expose the RTCs to competition but at the same time ensure a level playing field. These also need to be free of control if they have to face competition. One group alone cannot have unbridled freedom.

Privatization has to be thought segment wise. The segments prescribed for policymaking are as follows:

1. Interstate.
2. Interdistrict.
3. Intradistrict.
4. Urban.
5. Interior rural and hilly.

Each one of these segments would require a tailor-made approach and solution. Privatization also has to proceed keeping this segmentation in mind.

Of course, in all cases of participation by private parties, the governments should prescribe a minimum number of fleets that should be owned by them and corresponding infrastructure. It should specify quality parameters and labor requirements. It should also specify the maximum number of players who can operate in a segment. We suggest three to four players in each segment of which one should be a government operator. This helps to control if they tend to operate as oligopolists and will also help to set benchmarks. While privatizing routes, exercise has to be undertaken to assess the traffic needs. This should be in fact continuous exercise. It should avoid excessive competition or lack of service.

Small fleet operators can join together under a banner and compete. One can as such prescribe that each operator should have at least 50 buses which can be scaled up later. They can have it either on their own or come together. This will ensure some economy for them also. This will also ensure that they maintain a certain level of infrastructure. The policy should have sufficient entry barriers so that everybody cannot enter the field at will. It should also depend upon the due diligence of the operators. They should exhibit better management practices in terms of crew training, information systems and accounting, compliance, and engineering.

The licensing will be based on agreed-upon norms for fare fixation, traffic flow, and productivity, as it has been implemented in the highway sector.

Interstate routes are the most lucrative ones. These routes can be auctioned and the surplus from this can be put into a fund which can help subsidize operators who work in loss-making routes. They can be auctioned on the basis of rent which they are willing to pay for the quarter. It can be per bus or per schedule.

Intrastate or interdistrict are also fairly profit-making operations. Here also it can be auctioned. Alternatively, interdistrict and intradistrict can be combined as these run parallel in many segments. The government can form blocks of these routes and can auction the entire block. As mentioned above, each block can have three players of which one will be RTC. The blocks should be formed such that it should include also loss-making routes. The operators are expected to cross-subsidize and still return a surplus to the government.

Purely loss-making routes like interior rural routes or routes covering remote areas or hilly areas can be treated separately. These would form the universal service obligations. These can be auctioned on what kind of support would be needed. They can ask for tax waivers or for financial support. This is supposed to be met through generated by profit making routes.

In urban sectors, again some support may be needed. Here again they can form blocks and auction these. Here the operations may not be viable, as it will have huge concessional segments, low vehicle utilization, and off-peak operations.

Alternatively, each route can be auctioned. Here if a route is identified as profitable, it should be on the basis of license fees that an operator is willing to pay. If it is expected to be a loss-making route, then it should be based on the support that they would require. This principle is followed in London transport. Here again it will lead to proliferation of operators.

Another alternative which can be pursued is that RTCs should increasingly outsource their operations. They can hire buses instead of owning all the buses. They can easily hire up to 25% of their fleet strength. This will also meet the objective of involving private entrepreneurs in this sector. There are various practices for compensating these bus owners. Each RTC can device its own practice. It can also outsource maintenance activities which are non-critical. It can even outsource its existing maintenance infrastructure which has huge excess capacities. Operations and management can be left to private operators. This will also help in reducing costs. In fact, many states have their own tire rethreading units also.

Privatizing operations and allowing private parties will in fact lead to better tax realization as they are now operating illegally.

In privatization, globally, the experience has been that it is equally necessary for governments to ensure viability of both state-run corporations and private operators. Initial responses will be toward curtailing RTCs, but it should be taken care that it does not lead to sickness of RTCs. One suggestion that is made is that RTCs themselves can act as the licensing body. It can decide which routes are to be privatized and what are the license fees. It is felt this might lead to conflict of interest, and it may not lead to the desired result of aggressive privatization. Here RTC itself is a stakeholder, and, so, they may be viewed always suspiciously both by private operators and general public.

5.4 ASSESSMENT OF USER SATISFACTION

Commuters in any urban area are quite in their demands and perceptions. For example, commuters using their own vehicles for work trips will be very happy if the speed of operations even during peak periods is nearer to the design speed of the roads, enjoy free parking, and do not encounter any difficulty in finding a parking place at any point in time at all the places they visit. Cyclists will be very happy if they can enjoy an exclusive ROW and proper arrangements for easy maneuverability at junctions. Pedestrians very much desire to have adequate quality footpaths and opportunities for safe crossing facilities on all busy corridors. The city officials need to conduct periodic perception surveys of commuters using the facilities at least once a year. This data is very useful in developing appropriate improvement measures to maximize commuter perception.

There is another large section of commuters (about 50–60%) using mass public transportation systems (bus, suburban rail, and metro systems) in major metropolitan cities. There is an urgent necessity that their satisfaction levels be given serious importance in order to enhance the use of public mass transportation facilities. Primary surveys need to be undertaken to gauge their perceptions and their views on areas of improvement. The stakeholders belonging to this category could be broadly divided into three groups, namely:

1. Service providers of mass public transport facilities (employees, drivers, etc.).
2. Commuters using the facilities.
3. Opinion makers (knowledgeable individuals who can influence both service providers and commuters using the facilities).

The primary data collection schedules for obtaining the views are given in the Appendix of this chapter. The Indian Institute of Management is involved in collecting this primary data from about 60,000 persons in the state of Karnataka and Andhra Pradesh (combined). The findings of these primary surveys are used in developing appropriate strategies for enhancing the ridership of mass transportation facilities.

5.5 THE GLOBAL TREND

The policies of government have been going against global trends and local needs. The global trend is to treat public transport as a public utility and as a contributor to overall development. Public transport services, whether provided by state or private operators, are subsidized in most countries. For example, in European countries like the UK, France, Sweden, and Switzerland, the cost of operation of public transport buses is subsidized by the states from about 40% to 75%. In some European countries, there is a practice of imposing taxes on private vehicles or on firms employing more than a minimum number of people to raise resources for subsidizing public transport facilities as bus transport is treated more as a public utility service than as a commercial undertaking. The global trend is also to encourage public transport and discourage privately owned vehicles. In India, the policy appears to be working against both these propositions. It encourages privately owned vehicles by offering tax exemptions/reductions and is punitive in its taxation of public transport. Within public transport, it is more supportive of private than state-owned providers.

SUMMARY

This chapter mainly discusses the importance of problems, and efficient use of the public transport management infrastructure. It also highlights the improved vehicular flow at different conflict intervals, reversible lanes, channelized sections, and also of high-occupancy vehicles in vehicular and non-vehicular modes. It also gives information about the data collection and analysis for effective transport planning and sheds some light on the role of private sectors in some selected states and also says how government policies affect the global trend and user satisfaction.

Appendix

RESPONDENT CATEGORY

Public Transport Company Opinion Makers

Please circle the most appropriate single alternative, unless it is clearly specified that you can choose more than one alternative. Your responses will help us evolve means to enhance the functioning of PUBLIC TRANSPORT COMPANY. At no instance will your answers be identified with you.

1. Name of the interviewer:
2. Date of the interview:
3. Quality checks: Field Supervisor to circle and sign.
4. Guided interview, Training 1, Scrutiny 2, Back Check 3
5. Signature:___________ Name:_____________
6. ZONE
   Hyderabad 1, Telengana 2, Rayalaseema 3, Andhra 4
7. Location: (place of interview)
   Home/School/Office/Shop 1
   City/Town/Village name 2
   District:
8. Name of the respondent:
9. Address:

Instructions to Interviewers: Please record response as specified (Circle/Tick/Write/etc.) for each question. Use dark pencil or blue color ballpoint pen. Be courteous to respondents and interview only those people who wish to be part of the survey by answering all questions.

Questionnaire: Opinion Makers

1. Gender: (Circle response) Male 1, Female 2
2. Please tell me your age (in years): (Circle response)
   <20 **1**, TERMINATE >65 **5**, TERMINATE
   21–35 2, 36–50 3, 51–65 4
3. Residential status: (Circle response) Urban 1, Rural 2
4. Please tell me your educational qualification: (Circle response)
   Illiterate 1, Literate Primary School 2, High School 3
   Graduation 4, Post-Graduation 5
5. Please tell me your occupation. Specify designation for 3, 4, 8, 9, 11, and circle response code.
   1 Student
   2 Housewife
   3 Govt. Servant Central

4 Govt. Servant State
5 Business Person
6 Social Worker/NGO
7 Teacher
8 Community Leader
9 Private Company
10 Agriculture
11 Elected Representative
12 Others

6. Please tell me your average family monthly income (in rupees): (Circle response)
   <5000 1, 5001–10,000 2, 10,001–15,000 3, >15,000 4
7. Are you a bus pass holder? (Circle response) Yes 1 (Continue) No 2 (GO TO Q. 8)
   If Yes, specify category of PASS HOLDER:
   1) Concession Bus Pass: Student 1, Employee 2, Any other 3
   2) Non-Concession Bus Pass: Daily 1, Monthly 2
   3) Since when are you using the bus pass: (specify years/months)
8. Current number of bus pass holders in your family: (Circle response)
   0, 1, 2, 3, 4, More than **5**
9. Duration of experience as a bus passenger (in years): (Circle response)
   Less than 1, 2, 3, 4, 5, More than 5
10. As a bus passenger, the travel has been mostly in:
    (Circle ONE response only for each type of travel)
    1. City
    Ordinary **1**, Metro Express **2**, Metro Liner **3**, Veera **4**
    2. Intercity
    Ordinary **1**, Express **2**, Semi-luxury **3**, Hitech **4**, Volvo **5**
11. Do you own any of the following (tick as many as appropriate)?

Sl. No.	Ownership of Vehicle	Tick Your Response
1	2 -Wheeler	
2	Car	
3	Auto	
4	Taxi	
5	Maxicab/Matador/Trax/Minibus	
6	Bus	

12. Have you ever traveled in buses other than PUBLIC TRANSPORT COMPANY? (Circle response) Yes **1**, No **2**
13. People travel in PUBLIC TRANSPORT COMPANY buses because (tick as many as appropriate):

Sl. No.	Reasons to Travel in PUBLIC TRANSPORT COMPANY	Tick Your Opinion
1	They have no better alternative	
2	It is the safest	
3	It is the most comfortable	
4	It is the most convenient in terms of timing	
5	It is the cheapest	
6	It is run by a government corporation	
7	It is meant for common good of society	
8	It offers concessions to the needy	

14. Share your opinion on the following statement using the scale:
    1 = Strongly Disagree, **2** = Disagree, **3** = Uncertain/Neutral,
    4 = Agree, **5** = Strongly Agree

Sl. No.	Statement	Scale Agree/Disagree				
1	Public should be encouraged to use personalized vehicles than relying on PUBLIC TRANSPORT COMPANY	1	2	3	4	5
2	The condition of roads in my area are pathetic	1	2	3	4	5
3	There is a heavy misuse of bus passes	1	2	3	4	5
4	Smaller vehicles are better suited than normal buses	1	2	3	4	5
5	Because of shortage of buses from PUBLIC TRANSPORT COMPANY, people are forced to use personalized vehicles	1	2	3	4	5

15. Denote your preferences on the below given various travel attributes for PUBLIC TRANSPORT COMPANY as per the rating scale: Refer to scale and circle one response for each attribute.
    1 = Not at all Important, **2** = Less Important, **3** = Important,
    4 = Very Important, **5** = Most Important

Sl. No.	Statement	Rating Scale				
1	Time of the schedule	1	2	3	4	5
2	Fare (cost of the ticket)	1	2	3	4	5
3	Comfort in the bus	1	2	3	4	5
4	Facilities in bus shelter/stand/terminus	1	2	3	4	5
5	Reliability of service	1	2	3	4	5

(Continued)

(Continued)

Sl. No.	Statement	Rating Scale				
6	Safety of the journey	1	2	3	4	5
7	Travel duration (journey time)	1	2	3	4	5
	NON-PUBLIC TRANSPORT COMPANY travel attribute	**Rating Scale**				
8	Passenger friendly behavior of the staff	1	2	3	4	5
9	Balance of speed with safety	1	2	3	4	5
10	Balance of punctuality and customer needs	1	2	3	4	5

16. Denote your judgment regarding the quality of PUBLIC TRANSPORT COMPANY service on the various attributes given below, as per the rating scale:
Refer to scale and circle one response for each attribute.
1 = Very Poor, **2** = Poor, **3** = Average, **4** = Good, **5** = Very Good

Sl. No.	Quality of PUBLIC TRANSPORT COMPANY Service	Rating Scale				
1	Punctuality	1	2	3	4	5
2	Speed	1	2	3	4	5
3	Frequency	1	2	3	4	5
4	Seat availability	1	2	3	4	5
5	Fare (cost of the ticket)	1	2	3	4	5
6	Reliability	1	2	3	4	5
7	Convenient timings	1	2	3	4	5
8	Comfort in the bus	1	2	3	4	5
9	Facilities in bus shelter/stand/terminus	1	2	3	4	5
10	Number of stops during the journey	1	2	3	4	5
11	Handling of luggage	1	2	3	4	5

17. Describe the behavior of PUBLIC TRANSPORT COMPANY staff on the various attributes given below, as per the rating scale. Refer to scale and circle one response for each attribute.
1 = Very Poor, **2** = Poor, **3** = Average, **4** = Good, **5** = Very Good

Sl. No.	PUBLIC TRANSPORT COMPANY Staff Behavior Rating	Rating Scale				
1	Courtesy	1	2	3	4	5
2	Honesty: (returning small change)	1	2	3	4	5
3	Honesty: (diligently issuing tickets)	1	2	3	4	5
4	Helpful to: (women, elderly, children)	1	2	3	4	5
5	Discipline in stopping when required	1	2	3	4	5

18. Have you ever noticed breakdown of buses? Yes 1, No 2
(If the answer is Yes, continue; otherwise, go to question number 19).

18.1 The vehicles you noticed were mostly of:
1 PUBLIC TRANSPORT COMPANY
2 Non-PUBLIC TRANSPORT COMPANY
3 BOTH

19. Did any time the bus you were on traveling break down? Yes **1,** No **2**
(If the answer is Yes, continue; otherwise, go to question number 22)

20. The bus belonged to: 1 PUBLIC TRANSPORT COMPANY, **2** Non-PUBLIC TRANSPORT COMPANY, **3** BOTH
The reasons attributable for the breakdown are (Tick as many as appropriate):

Sl. No.	Reasons for Bus Breakdown: Tick Your Opinion
1	Poor maintenance of vehicle
2	Bad conditions of road
3	Aged vehicle
4	Rude handling by driver
5	Any other

21. Describe your overall post-breakdown experience as:

Sl. No.	Post-Breakdown Experience: Tick Your Opinion
1	Best available alternative was arranged quickly
2	Taken care of well, until the alternatives were made available
3	Left to ourselves/not at all cared for

22. Compare PUBLIC TRANSPORT COMPANY with private buses (Non-PUBLIC TRANSPORT COMPANY) on the following various parameters as per the rating scale. Refer to scale and circle one response for each attribute.
1 = Worse than private, **2** = Equal to private,
3 = Slightly better than private,
4 = Much better than private, **5** = Excellent

Sl. No.	Service Comparison PUBLIC TRANSPORT COMPANY versus Private Operators	Circle Response
1	Timing of the schedule	1 2 3 4 5
2	Fare (cost of the ticket)	1 2 3 4 5
3	Comfort in the bus	1 2 3 4 5
4	Facilities in bus shelter/stand/terminus	1 2 3 4 5
5	Reliability of service	1 2 3 4 5
6	Safety of the journey	1 2 3 4 5

(Continued)

(Continued)

Sl. No.	Service Comparison PUBLIC TRANSPORT COMPANY versus Private Operators	Circle Response				
7	Travel duration	1	2	3	4	5
8	Passenger-friendly behavior of the staff	1	2	3	4	5
9	Balance of speed with safety	1	2	3	4	5
10	Balance of punctuality and customer needs	1	2	3	4	5

23. List the best practices of private (Non-PUBLIC TRANSPORT COMPANY) buses, which could be emulated by PUBLIC TRANSPORT COMPANY. (Please tick as many as appropriate):
Best Service Quality by PUBLIC TRANSPORT COMPANY **1**,
Best Service Quality by NON-PUBLIC TRANSPORT COMPANY **2,**
Both Are Same **3**

		Service Required		
Sl. No.	Service Quality	PUBLIC TRANSPORT COMPANY	Non-PUBLIC TRANSPORT COMPANY	Both are Same
1	Comfortable seating	1	2	3
2	Leg room between seats	1	2	3
3	Convenient places for boarding	1	2	3
4	Convenient places for alighting	1	2	3
5	Convenient stops during the journey for snacks/coffee/lunch/dinner	1	2	3
6	Passenger need-based stops for relaxation and attending toilet requirements	1	2	3
7	Better quality entertainment like music/ video	1	2	3
8	Easy ticketing facility	1	2	3
9	Cost effectiveness	1	2	3
10	Less travel time	1	2	3
11	Convenient departure and arrival timings	1	2	3
12	Cleanliness of exterior	1	2	3
13	Easy placement of luggage	1	2	3
14	Less rash driving	1	2	3
15	Empathic behavior by staff	1	2	3

24. To make PUBLIC TRANSPORT COMPANY continue to provide efficient and cost effective transport service, it should enjoy functional freedom and autonomy in: (Please tick as many as appropriate.)
Functional Freedom Tick Your Opinion
1 Increasing fares for reasons beyond its control like hike in diesel price/ input costs
2 Deciding the routes and their frequencies

3 Having differentiated services for various customer segments
4 Deciding different fares for off-peak and other special services
5 Cancellation of schedules/trips/services when there is no patronage

25. If there is a hike in fare, I shall: (Circle any one response only)
Fare Hike My opinion (Circle any one response only)
1 Grudgingly continue to travel in PUBLIC TRANSPORT COMPANY
2 As a responsible citizen, empathize with PUBLIC TRANSPORT COMPANY and continue to travel
3 Go for alternative modes
4 Join agitation if someone gives a call

26. Share your impressions about PUBLIC TRANSPORT COMPANY, by using the scale: (Circle any one response)
1 = Strongly Disagree, **2** = Disagree, **3** = Uncertain/Neutral,
4 = Agree, **5** = Strongly Agree

Sl. No.	Your Impressions about PUBLIC TRANSPORT COMPANY	Scale Agree/Disagree				
1	Is providing best services to public	1	2	3	4	5
2	Needs support from government to do better	1	2	3	4	5
3	Should be privatized	1	2	3	4	5
4	Has grown too big, hence needs to be bifurcated	1	2	3	4	5
5	Should be freed from governmental control	1	2	3	4	5
6	Whatever is done, still there is always shortage of buses experienced by public	1	2	3	4	5

27. Is there a need to increase number of services in your area? Yes **1,** No **2.** If Yes: Could you provide details of starting point and ending point?
Route ________ From ________ To
Route 1 ________________
Route 2 ________________
Route 3 ________________

28. Is there a need to decrease number of services in your area? Yes **1,** No **2.** If Yes: Could you provide details of starting point and ending point?
Route ________ From ________ To
Route 1 ________________
Route 2 ________________
Route 3 ________________

Thank you!

Stakeholders Opinion Survey Format for Commuters

Please circle the most appropriate single alternative, unless it is clearly specified that you can choose more than one alternative. Your responses will help us evolve means to enhance the functioning of Bus Company. At no instance will your answers be identified with you.

Name of the interviewer: ———— Date of the interview: ————————
2005 Quality checks: Field Supervisor to circle and sign. ————
Guided interview: Training 1, Scrutiny 2, Back Check 3
Signature: ———— Name: ————
Name of the District:————————
Location: (Place of interview)————
On board (fm———— to————) 1
Bus Stand———— 2
(City/Town/Village name) ————
District:————
Name of the passenger (respondent):————
Address:————————

Instructions to Interviewers:

Please record response as specified (Circle/Tick/Write/etc.) for each question. Use dark pencil or blue color ballpoint pen. Be courteous to respondents and interview only those people who wish to be part of the survey by answering all questions. Do not interview any person without obtaining their full consent for the interview.

Questionnaire: Public Transport Company Commuters

1. Gender: (Circle response) Male 1, Female 2
2. Please tell me your age (in years): (Circle response)
   <20 **1** TERMINATE, >65 **5** TERMINATE
   21–35 **2**, 36–50 3, 51–65 **4**
3. Residential status: (Circle response) Urban 1, Rural 2
4. Please tell me your educational qualification: (Circle response)
   Illiterate **1**, Literate Primary School **2**, High School **3**
   Graduation **4**, Post-Graduation **5**
5. Please tell me your occupation: (Circle response)
   Student **1**, Housewife **2**, Govt. Servant **3**, Business Person **4**, Others **5**
6. Please tell me your average family monthly income (in rupees): (Circle response)
   <5000 **1**, 5001–10,000 **2**, 10,001–15,000 **3**, >15,000 **4**
7. Are you a bus pass holder? (Circle response) Yes 1 (Continue), No 2 (Go to Q. 8)
   If Yes, specify category of PASS HOLDER:
   1) Concession bus pass: Student 1, Employee 2, Any other 3
   2) Non-Concession bus pass: Daily 1, Monthly 2
   3) Since when are you using the bus pass? ————————(specify years/months)
8. Current number of bus pass holders in your family: (Circle response)
   0, 1, 2, 3, 4, More than **5**
9. Duration of experience as a bus passenger (in years): (Circle response)
   <1, 2, 3, 4, 5, >5

10. As a bus passenger, the travel has been mostly in:
    (Circle ONE response only for each type of travel)
    1. City
       Ordinary **1**, Metro Express **2**, Metro Liner **3**, Veera **4**
    2. Intercity
       Ordinary **1,** Express **2,** Semi-luxury **3**, Hitech 4, Volvo **5**
11. Have you ever traveled in buses other than PUBLIC TRANSPORT COMPANY? (Circle response) Yes 1 No 2
12. People travel in PUBLIC TRANSPORT COMPANY buses because (Tick as many as appropriate):

Sl. No.	Reasons to travel in PUBLIC TRANSPORT COMPANY	Tick Your Opinion
1	They have no better alternative	
2	It is the safest	
3	It is the most comfortable	
4	It is the most convenient in terms of timing	
5	It is the cheapest	
6	It is run by a government corporation	
7	It is meant for common good of society	
8	It offers concessions to the needy	

13. Compare PUBLIC TRANSPORT COMPANY with private buses (Non-PUBLIC TRANSPORT COMPANY) on the following various parameters as per the rating scale. Refer to scale and circle one response for each attribute.
    1 = Worse than private, **2** = Equal to private,
    3 = Slightly better than private,
    4 = Much better than private, **5** = Excellent

Sl. No.	Service Comparison PUBLIC TRANSPORT COMPANY versus Private Operators	Circle Response
1	Timing of the schedule	1 2 3 4 5
2	Fare (cost of the ticket)	1 2 3 4 5
3	Comfort in the bus	1 2 3 4 5
4	Facilities in bus shelter/stand/terminus	1 2 3 4 5
5	Reliability of service	1 2 3 4 5
6	Safety of the journey	1 2 3 4 5
7	Travel duration	1 2 3 4 5
8	Passenger-friendly behavior of the staff	1 2 3 4 5
9	Balance of speed with safety	1 2 3 4 5
10	Balance of punctuality and customer needs	1 2 3 4 5

14. Denote your judgment regarding the quality of PUBLIC TRANSPORT COMPANY service on the various attributes given below, as per the rating scale: Refer to scale and circle one response for each attribute.
1 = Very Poor, **2** = Poor, **3** = Average, **4** = Good, **5** = Very Good

Sl. No.	Quality of PUBLIC TRANSPORT COMPANY Service	Rating Scale
1	Punctuality	1 2 3 4 5
2	Speed	1 2 3 4 5
3	Frequency	1 2 3 4 5
4	Seat Availability	1 2 3 4 5
5	Fare (cost of the ticket)	1 2 3 4 5
6	Reliability	1 2 3 4 5
7	Convenient timings	1 2 3 4 5
8	Comfort in the bus	1 2 3 4 5
9	Facilities in bus shelter/stand/terminus	1 2 3 4 5
10	Number of stops during the journey	1 2 3 4 5
11	Handling of luggage	1 2 3 4 5

15. Describe the behavior of PUBLIC TRANSPORT COMPANY staff on the various attributes given below, as per the rating scale. Refer to scale and circle one response for each attribute.
1 = Very Poor; **2** = Poor; **3** = Average; **4** = Good; **5** = Very Good

Sl. No.	PUBLIC TRANSPORT COMPANY Staff Behavior Rating	Rating Scale
1	Courtesy	1 2 3 4 5
2	Honesty: (returning small change)	1 2 3 4 5
3	Honesty: (diligently issuing tickets)	1 2 3 4 5
4	Helpful to: (women, elderly, children)	1 2 3 4 5
5	Discipline in stopping when required	1 2 3 4 5

16. Have you ever noticed breakdown of buses? Yes, No
(If the answer is Yes, continue; otherwise, go to question number 19).
17. If Yes, the vehicles you have noticed were mostly: **1**—of PUBLIC TRANSPORT COMPANY, **2**—of Non-PUBLIC TRANSPORT COMPANY
18. Did at any time the bus you were traveling on breakdown? Yes, No
(If the answer is Yes continue; otherwise, go to question number 22)
19. The bus belonged to: **1**—PUBLIC TRANSPORT COMPANY, **2**—Non-PUBLIC TRANSPORT COMPANY
The reasons attributable for the breakdown are (Tick as many as appropriate):
Reasons for bus breakdown—Tick Your Opinion

1	Poor maintenance of vehicle
2	Bad conditions of road
3	Aged vehicle
4	Rude handling by driver
5	Any other: __

20. Describe the overall post breakdown experience as:

Post-breakdown experience __ Tick Your Opinion	
1	Best available alternative was arranged quickly
2	Taken care of well, until the alternatives were made available
3	Left to ourselves/not at all cared for

21. Compare PUBLIC TRANSPORT COMPANY with private buses (Non-PUBLIC TRANSPORT COMPANY) on the following various parameters as per the rating scale. Refer to scale and circle one response for each attribute.
1 = Worse than Private, **2** = Equal to Private,
3 = Slightly Better than Private,
4 = Much Better than Private, **5** = Excellent

Sl. No.	Service Comparison PUBLIC TRANSPORT COMPANY versus Private Operators	Circle Response
1	Timing of the schedule	1 2 3 4 5
2	Fare (cost of the ticket)	1 2 3 4 5
3	Comfort in the bus	1 2 3 4 5
4	Facilities in bus shelter/stand/terminus	1 2 3 4 5
5	Reliability of service	1 2 3 4 5
6	Safety of the journey	1 2 3 4 5
7	Travel duration	1 2 3 4 5
8	Passenger friendly behavior of the staff	1 2 3 4 5
9	Balance of speed with safety	1 2 3 4 5
10	Balance of punctuality and customer needs	1 2 3 4 5

22. List the best practices of private (non-PUBLIC TRANSPORT COMPANY) buses, which could be emulated by PUBLIC TRANSPORT COMPANY. (Please tick as many as appropriate.)
Best Service Quality by PUBLIC TRANSPORT COMPANY 1
Best Service Quality by NON-PUBLIC TRANSPORT COMPANY 2
Both Are Same 3

		Service Required		
Sl. No.	Service Quality	PUBLIC TRANSPORT COMPANY	Non-PUBLIC TRANSPORT COMPANY	Both Are Same
1	Comfortable seating	1	2	3
2	Leg room between seats	1	2	3
3	Convenient places for boarding	1	2	3
4	Convenient places for alighting	1	2	3
5	Convenient stops during the journey for snacks/coffee/ lunch/dinner	1	2	3
6	Passenger need-based stops for relaxation and attending toilet requirements	1	2	3
7	Better quality entertainment like music/video	1	2	3
8	Easy ticketing facility	1	2	3
9	Cost effectiveness	1	2	3
10	Less travel time	1	2	3
11	Convenient departure and arrival timings	1	2	3
12	Cleanliness of exterior	1	2	3
13	Easy placement of luggage	1	2	3
14	Less rash driving	1	2	3
15	Empathic behavior by staff	1	2	3

23. To make PUBLIC TRANSPORT COMPANY continue to provide efficient and cost effective transport service, it should enjoy functional freedom and autonomy in: (Please tick as many as appropriate.)
Functional Freedom ___________ Tick Your Opinion
1 Increasing fares for reasons beyond its control like hike in diesel price/ input costs
2 Deciding the routes and their frequencies

3 Having differentiated services for various customer segments
4 Deciding different fares for off-peak and other special services
5 Cancellation of schedules/trips/services when there is no patronage

24. If there is a hike in fare, I shall: (Circle any one response only.)
Fare Hike, My opinion ___________ (Circle any one response only.)
1 Grudgingly continue to travel in PUBLIC TRANSPORT COMPANY
2 As a responsible citizen, empathize with PUBLIC TRANSPORT COMPANY and continue to travel
3 Go for alternative modes
4 Join agitation if someone gives a call

25. Share your impressions about PUBLIC TRANSPORT COMPANY, by using the scale: (Circle any one response.)
1 = Strongly Disagree, **2** = Disagree, **3** = Uncertain/Neutral,
4 = Agree, **5** = Strongly Agree

Sl. No.	Your Impressions about PUBLIC TRANSPORT COMPANY	Scale Agree/ Disagree
1	Is providing best services to public	1 2 3 4 5
2	Needs support from government to do better	1 2 3 4 5
3	Should be privatized	1 2 3 4 5
4	Has grown too big, hence needs to be bifurcated	1 2 3 4 5
5	Should be freed from governmental control	1 2 3 4 5
6	Whatever is done, still there is always shortage of buses experienced by public	1 2 3 4 5

26. Is there a need to increase number of services in your area? Yes 1, No 2. If Yes: Could you provide details of starting point and ending point?
Route _________ From _________ To _________ and
Route 1 __________________
Route 2 __________________
Route 3 __________________

27. Is there a need to decrease number of services in your area? Yes 1, No 2. If Yes: Could you provide details of starting point and ending point?
Route _________ From _________ To
Route 1 __________________
Route 2 __________________
Route 3 __________________

Thank you!

Respondent Category: Public Transport Company Employees

Please circle the most appropriate single alternative, unless it is clearly specified that you can choose more than one alternative. Your responses will help us evolve means to enhance the functioning of PUBLIC TRANSPORT COMPANY. At no instance will your answers be identified with you.

Name of the interviewer: ___________ Date of the interview:
Quality checks: Field Supervisor to circle and sign.
Guided interview: training 1, Scrutiny 2, Back Check 3
Signature: ___________ Name: ___________ Name of the District: ___________

Location: (Place of interview) On board (fm ___________ to ___________) 1
Bus Stand ___________ 2 (City/Town/Village name)
District: ___________
Name of the employee (respondent): _________________ Badge No.______________
Designation:
Driver 1, Conductor 2, Supervisor 3, Admin. Staff 4, Officer 5
Address: ______________

Instructions to Interviewers:

Please record response as specified (Circle/Tick/Write/etc.) for each question. Use dark pencil or blue color ballpoint pen. Be courteous to respondents and interview only those people who wish to be part of the survey by answering all questions. Do not interview any person without obtaining their full consent for the interview.

Questionnaire: Public Transport Company Employees

1. Gender: (Circle response) Male 1, Female 2
2. Please tell me your age (in years): (Circle response)
   <20 1 TERMINATE, >65 5 TERMINATE, 21–35 2, 36–50 3, 51–65 4
3. Residential status: (Circle response) Urban 1, Rural 2
4. Please tell me your educational qualification: (Circle response) Illiterate 1, Literate Primary School 2, High School 3, Graduation 4, Post-Graduation 5
5. Please tell me your average family monthly income (in rupees): (Circle response) <7000 =1, 7001–15,000 = 2, 15,001–25,000= 3, >25,000 = 4
6. Work experience (write response in no. of years): Total _________
   At PUBLIC TRANSPORT COMPANY at the current position: _________
7. Please tell me the number of promotions received so far during your tenure at PUBLIC TRANSPORT COMPANY.
   (Circle response) 0, 1, 2, 3, 4, >5
8. Have you ever traveled in buses other than PUBLIC TRANSPORT COMPANY? Yes, No
9. Do you own a vehicle? Yes 1 (Continue), No 2 (Go To Q.10)

9.(1) If yes, please specify the type of vehicle owned by you.
Two-wheeler = 1, Car = 2, Auto = 3, Maxi cab = 4,
Matador = 5, Minibus = 6, Tempo-Trax = 7, Lorry = 8, Bus = 9.

10. Please use the 5-point scale to answer this question: Circle ONE response only.
1 Strongly Disagree, 2 Disagree, 3 Uncertain, 4 Agree, 5 Strongly Agree
To make PUBLIC TRANSPORT COMPANY provide better service to people, in terms of its governance, it should be:

Sl. No.	Your Impressions about PUBLIC TRANSPORT COMPANY	Scale Agree/ Disagree				
1	Completely privatized	1	2	3	4	5
2	Partially privatized	1	2	3	4	5
3	Better controlled by state government	1	2	3	4	5
4	Enjoy better patronage by state government	1	2	3	4	5
5	Least control and interference by state government	1	2	3	4	5
6	Given unconditional autonomy	1	2	3	4	5
7	Ban completely the services of private operators, thus providing monopoly and larger scale of operations	1	2	3	4	5

11. Please use the 5-point scale to answer this question: Circle ONE response only.
1 Strongly Disagree, **2** Disagree, **3** Uncertain, **4** Agree, **5** Strongly Agree
The best option to help PUBLIC TRANSPORT COMPANY come out of the present financial crisis is to:

Sl. No.	Options To Solve the Financial Crisis at PUBLIC TRANSPORT COMPANY Scale	Scale Agree/ Disagree				
1	Allow PUBLIC TRANSPORT COMPANY to freely borrow from lending organizations	1	2	3	4	5
2	Sanction subsidies more liberally to shield from the impact of fuel hikes and concessional passes	1	2	3	4	5
3	Pump in more funds in the form of government equity with conditions of performance	1	2	3	4	5
4	Closing of some DEPOTS and/or merging with adjacent depots	1	2	3	4	5

12. Please use the 5-point scale to answer this question: Circle ONE response only.
1 Strongly Disagree, **2** Disagree, **3** Uncertain, **4** Agree, **5** Strongly Agree
Please specify your opinion on the major drawback of PUBLIC TRANSPORT COMPANY.

Sl. No.	Major Drawback of PUBLIC TRANSPORT COMPANY SCALE	Scale Agree/Disagree				
1	Government interference	1	2	3	4	5
2	Absence of professionalism at the policymaking level (Board)	1	2	3	4	5

(Continued)

(Continued)

Sl. No.	Major Drawback of PUBLIC TRANSPORT COMPANY SCALE	Scale Agree/Disagree				
3	Forced to offer services at rates lower than cost	1	2	3	4	5
4	Functioning of private operators	1	2	3	4	5
5	Lack of control on illegal operations	1	2	3	4	5

13. Please use the 5-point scale to answer this question: Circle ONE response only.
1 Strongly Disagree, **2** Disagree, **3** Uncertain, **4** Agree, **5** Strongly Agree
People travel in PUBLIC TRANSPORT COMPANY bus because:

Sl. No.	Reasons to Travel in PUBLIC TRANSPORT COMPANY Bus	Scale Agree/Disagree				
1	They have no better alternative	1	2	3	4	5
2	It is the safest	1	2	3	4	5
3	It is the cheapest	1	2	3	4	5
4	It is run by government	1	2	3	4	5
5	It is meant for common good of society	1	2	3	4	5

14. Please use the 5-point scale to answer this question: Circle ONE response only.
1 Strongly Disagree, **2** Disagree, **3** Uncertain, **4** Agree, **5** Strongly Agree
Share your agreement/disagreement on public transport opinion statements

Sl. No.	Opinion on Public Transport	Scale Agree/ Disagree				
1	Public should be encouraged to use two-/ four-wheeler rather than relying on PUBLIC TRANSPORT COMPANY	1	2	3	4	5
2	The conditions of roads in my area are pathetic	1	2	3	4	5
3	Until roads are improved, public should use their own transport rather than asking for PUBLIC TRANSPORT COMPANY buses	1	2	3	4	5
4	There is a heavy misuse of bus passes	1	2	3	4	5
5	Smaller vehicles are better suited than normal buses	1	2	3	4	5

15. Please use the 5-point scale to answer this question: Circle ONE response only.
1 = Not at all important, **2** = Less important, **3** = Important, **4** = Very important, **5** = Most important
Denote your preferences on the below given travel attribute as per the rating scale.

Sl. No.	Service Comparison PUBLIC TRANSPORT COMPANY versus Private Operators	Circle Response				
1	Timing of the schedule	1	2	3	4	5
2	Fare (cost of the ticket)	1	2	3	4	5
3	Comfort in the bus	1	2	3	4	5
4	Facilities in bus shelter/stand/ terminus	1	2	3	4	5
5	Reliability of service	1	2	3	4	5
6	Safety of the journey	1	2	3	4	5
7	Travel duration	1	2	3	4	5
8	Passenger friendly behavior of the staff	1	2	3	4	5
9	Balance of speed with safety	1	2	3	4	5

16. Denote your judgment regarding the quality of PUBLIC TRANSPORT COMPANY service on the various attributes given below, as per the rating scale. Refer to scale and circle one response for each attribute.
1 = Very Poor, **2** = Poor, **3** = Average, **4** = Good, **5** = Very Good

Sl. No.	Quality of PUBLIC TRANSPORT COMPANY Service	Rating Scale				
1	Punctuality	1	2	3	4	5
2	Speed	1	2	3	4	5
3	Frequency	1	2	3	4	5
4	Seat availability	1	2	3	4	5
5	Fare (cost of the ticket)	1	2	3	4	5
6	Reliability	1	2	3	4	5
7	Convenient timings	1	2	3	4	5
8	Comfort in the bus	1	2	3	4	5
9	Facilities in bus shelter/ stand/terminus	1	2	3	4	55
10	Number of stops during the journey	1	2	3	4	5
11	Handling of luggage	1	2	3	4	5

17. Describe the behavior of PUBLIC TRANSPORT COMPANY staff on the various attributes given below, as per the rating scale. Refer to scale and circle one response for each attribute.
1 = Very Poor, **2** = Poor, **3** = Average, **4** = Good, **5** = Very Good

Sl. No.	PUBLIC TRANSPORT COMPANY Staff Behavior Rating	Rating Scale				
1	Courtesy	1	2	3	4	5
2	Honesty: (returning small change)	1	2	3	4	5
3	Honesty: (diligently issuing tickets)	1	2	3	4	5
4	Helpful to: (women, elderly, children)	1	2	3	4	5
5	Discipline in stopping when required	1	2	3	4	5

18. Have you ever experienced a breakdown of buses en route when you TRAVELED IN PUBLIC TRANSPORT COMPANY bus?
    Yes 1 (CONTINUE), No 2 (GO TO Q. 20)

18.1 If Yes, the reasons attributed for it are:
    Please use the 5-point scale to answer this question: Circle ONE response only.
    1 Strongly Disagree, 2 Disagree, 3 Uncertain, 4 Agree, 5 Strongly Agree

Sl. No.	Reasons for Breakdown	Rating Scale				
1	Poor maintenance (of vehicle)	1	2	3	4	5
2	Bad conditions of road	1	2	3	4	5
3	Aged vehicle	1	2	3	4	5
4	Rude handling by driver	1	2	3	4	5
5	Any other	1	2	3	4	5

19. How was the post-breakdown experience?
    Best available alternative was arranged quickly—Yes 1, No 2
    Taken care of well until the alternatives were made available—Yes 1, No 2
    We were left to ourselves, no one helped us—Yes 1, No 2
20. Compare PUBLIC TRANSPORT COMPANY with private buses (non-PUBLIC TRANSPORT COMPANY) on the following parameters as per the rating scale.
    1 = Worse than Private, **2** = Equal to Private,
    3 = Slightly Better than Private,
    4 = Much Better than Private, **5** = Excellent

Sl. No.	Service Comparison PUBLIC TRANSPORT COMPANY versus Private Operators	Circle Response				
1	Timing of the bus	1	2	3	4	5
2	Fare (cost of the ticket)	1	2	3	4	5
3	Comfort in the bus	1	2	3	4	5
4	Facilities in bus shelter/stand/terminus	1	2	3	4	5

(Continued)

(Continued)

Sl. No.	Service Comparison PUBLIC TRANSPORT COMPANY versus Private Operators	Circle Response				
5	Reliability of service	1	2	3	4	5
6	Safety of the journey	1	2	3	4	5
7	Travel duration	1	2	3	4	5
8	Passenger-friendly behavior of the staff	1	2	3	4	5
9	Balance of speed with safety	1	2	3	4	5
10	Balance of punctuality and customer needs	1	2	3	4	5

21. List the best practices of Private (Non-PUBLIC TRANSPORT COMPANY) buses, which could be emulated by PUBLIC TRANSPORT COMPANY: (Please tick as many as appropriate.)
Best Service Quality by PUBLIC TRANSPORT COMPANY 1,
Best Service Quality by NON-PUBLIC TRANSPORT COMPANY 2,
Both are Same 3

		Service Required		
Sl. No.	Service Quality	PUBLIC TRANSPORT COMPANY	Non-PUBLIC TRANSPORT COMPANY	Both are Same
1	Comfortable seating	1	2	3
2	Leg room between seats	1	2	3
3	Convenient places for boarding	1	2	3
4	Convenient places for alighting	1	2	3
5	Convenient stops during the journey for snacks/coffee/lunch/dinner	1	2	3
6	Passenger need-based stops for relaxation and attending toilet requirements	1	2	3
7	Better quality entertainment like music/video	1	2	3
8	Easy ticketing facility	1	2	3
9	Cost effectiveness	1	2	3
10	Less travel time	1	2	3
11	Convenient departure and arrival timings	1	2	3
12	Cleanliness of interior	1	2	3
13	Cleanliness of exterior	1	2	3
14	Easy placement of luggage	1	2	3
15	Less rash driving	1	2	3
16	Empathic behavior by staff	1	2	3

22. To make PUBLIC TRANSPORT COMPANY continue to provide efficient and cost-effective transport service, it should enjoy freedom and autonomy: (Please tick as many as appropriate.)

    SCOPE OF FREEDOM and AUTONOMY Tick Your Opinion

    1 In increasing fares for reasons beyond its control like diesel price increase ________

    2 In deciding the routes and their frequencies ________

    3 In having differentiated services for various customer segments ________

    4 In deciding different fares for differentiated services ________

    5 In managing its own finances and assets ________

    6 In evolving its own policies and procedures in enhancing customer satisfaction ________

25. Share your impressions about PUBLIC TRANSPORT COMPANY, by using the scale: (Circle any one response.)

    1 = Strongly Disagree, **2** = Disagree, **3** = Uncertain/Neutral,
    4 = Agree, **5** = Strongly Agree

Sl. No.	Your Impressions about PUBLIC TRANSPORT COMPANY	Scale Agree/Disagree				
1	Is providing best services to public	1	2	3	4	5
2	Needs support from government to do better	1	2	3	4	5
3	Should be privatized	1	2	3	4	5
4	Has grown too big, hence needs to be bifurcated	1	2	3	4	5
5	Should be freed from governmental control	1	2	3	4	5
6	Whatever is done, still there is always shortage of buses experienced by public	1	2	3	4	5

26. Is there a need to increase number of services in your area? Yes 1, No 2. If Yes: Could you provide details of starting point and ending point?

    Route ________ From ________ To ________

    Route 1 ________

    Route 2 ________

    Route 3 ________

27. Is there a need to decrease number of services in your area? Yes 1, No 2. If Yes: Could you provide details of starting point and ending point?

    Route ________ From ________ To ________

    Route 1 ________

    Route 2 ________

    Route 3 ________

Thank you!

6 Special Focus on Public Mass Transport Planning in Rural Areas

6.1 RURAL TRANSPORT SCENARIO CHANGES AND THEIR IMPACT ON ECONOMIC DEVELOPMENT AND DEMAND FOR IMPROVED TRANSPORT

About 75% of India's population lives in small villages with a few hundred people in a settlement. These villages normally follow caste-based clusters, an old social class system that is not exhibited in most urban areas. The forward caste people belonging to the top of the caste system live in the central part of the village and the people of the backward caste live on the outskirts.

Most forward caste people are rich and better educated compared to people belonging to backward caste. Figure 6.1 depicts a typical village setting in India.

The information on the growth of population in rural settlements is presented in Table 6.1; it may be noted that for settlements with population less than 2000 (four settlement sizes), the growth rates are either negative or very small. Settlements sizes 5000–10,000 and above generate higher growth rates.

The population of India is distributed in 580,000 settlements of different categories spread across the country. For the purpose of planning, the settlements are divided into "seven" categories based on the population size of the settlements. Accessibility is considered a prerequisite for economic development. However, at the turn of this century, almost four decades after independence, most of the rural Indian settlements were still not linked with all-weather roads (AWRs). The initiation of Pradhan Mantri Gram Sadak Yojana (PMGSY) during the year 2000 has provided impetus for providing AWRs to different categories of rural segments. Rural road connectivity is not only a key component of rural development by promoting access to economic and social services and generating increased agricultural incomes and productive employment opportunities in India but also a key ingredient in ensuring sustainable poverty reduction. Due to the efforts made, over the years, at the state and central levels, through different programs, most of the habitations (with population greater than 500) in the country are already connected by AWRs (Figure 6.2).

6.2 PMGSY (DEDICATED RURAL TRANSPORT DEVELOPMENT)

The PMGSY was launched by the Government of India as part of its strategy to reduce poverty. This centrally sponsored program aims to establish AWR connectivity for

DOI: 10.1201/ 9781003740247-6

FIGURE 6.1 A Typical Village in India

TABLE 6.1
Distribution of Population in Rural Settlements

	Rural (Population Class in Million)		
Population Classes	**1991**	**2001**	**2011**
Less than 200	10.52	9.2	8.17
200–499	48.45	43.96	39.68
500–999	104.38	105.3	103.32
1000–1999	160.31	183.35	197.53
2000–4999	185.59	239.19	288.77
5000 [illegible]	69.81	98.1	123.88
>10,000	43.721	63.49	[illegible]
Total	622.81	742.62	833.72

Source: Government of India Ministry of Statistics and Programme Implementation Central Statistics Office Social Statistics Division. Census 2001 and 2011.

rural areas that remain unconnected. In plain regions, the scheme focuses on connecting habitations with populations of 500 or more, while in special areas such as hilly regions, deserts, tribal zones, and designated backward districts, it targets habitations with populations of 250 or more.

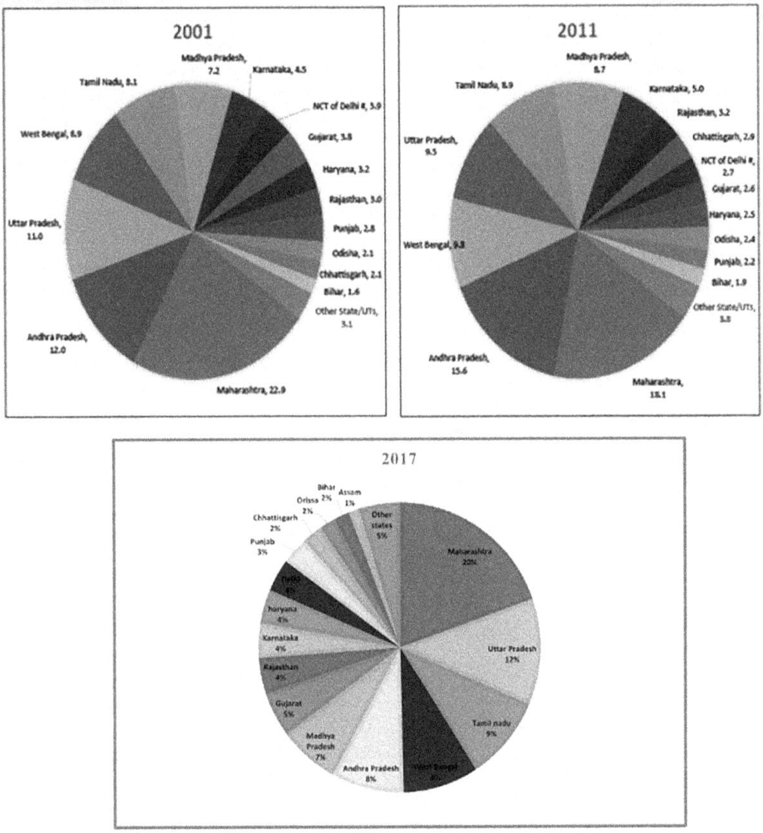

FIGURE 6.2 State Share of Slum Population to Total Slum Population of India

Source: Primary census abstract for slum 2011, Office of the Registrar General and Census Commissioner, New Delhi India; Report of the Committee on Slum Statistics/Census, Ministry of Housing and Urban Poverty Alleviation, New Delhi, India

Out of a total of 178,184 eligible habitations (populations of 250+ and 500+), 16,086 habitations were connected through state resources, and 4952 were either excluded or found unsuitable for the program. Of the remaining 157,146 habitations approved under PMGSY, 156,617 had been connected by March 31, 2024, leaving 529 habitations yet to be completely connected. For habitations with populations between 100 and 249 in Left Wing Extremism (LWE)-affected areas, 6245 were approved for road connectivity, out of which 6034 were connected by the same date.

Under PMGSY-I, a total of 644,878 km of road length was approved for new road connectivity and upgrades, with 624,187 km completed by March 31, 2024. To further enhance the efficiency of the rural road network and support socioeconomic growth, PMGSY-II was launched in 2013. This phase aimed at upgrading selected Through Routes and Major Rural Links (MRLs) with a target of 50,000 km. By March 31, 2024, 48,962 km of the sanctioned 49,857 km had been completed.

In 2019, PMGSY-III was introduced to strengthen and consolidate 125,000 km of Through Routes and Major Rural Links, with the aim of connecting habitations to essential services such as agricultural markets, secondary schools, and hospitals. As of March 31, 2024, 115,408 km had been sanctioned across 29 states, with 78,346 km completed. This phase is scheduled for completion by March 2025. Table 6.2 and Figure 6.3 give the information of the yearly achievement of PMGSY.

TABLE 6.2
Yearly Sanction Details of PMGSY

Sl. No.	Year	Total Number of Package	No. of Roads	No. of Habitation Covered	Pavements	
					Length (km)	Cost (Rs. in Million)
1	2000–01	2719	13,272	13,271	12,342	296,695.6
2	2001–02	3226	10,557	10,557	14,177	318,087.57
3	2003–04	3095	8478	13,413	30,124.05	46,348.3
4	2004–05	2885	7651	11,838	11,838	291,718.3
5	2005–06	4956	11,713	23,364	49,986.55	101,235.1
6	2006–07	8015	15,171	31,541	72,231.25	168,342.1
7	2007–08	9586	15,988	33,762	76,545.98	208,634.1
8	2008–09	12,881	18,730	24,892	82,660.65	246,622.8
9	2009–10	1912	3123	6400	14,905	39,792.8
10	2010–11	[illegible]	1070	1141	3047.21	8568.1
11	2011–12	4420	7506	10,590	25,480.84	73,934.6
12	2012–13	9082	14,216	25,052	48,738.779	169,579.7
13	2013–14	11,350	13,173	27,384	41,876.209	1,763,737.9
14	2014–15	672	829	1624	3898.822	133,021.72
15	2015–16	695	1653	1749	4300.531	163,834.49
16	2016–17	10,220	11,219	11,961	44,607.144	1,679,836.19
17	2017–18	9407	9100	15,600	34,154.370	1,394,914.99
18	2018–19	5830	5307	9764	31,166.808	1,356,887.76
19	2019–20	1312	1336	6246	10,441.830	555,993.40
20	2020–21	5763	5792	81,680	43,713.151	2,116,633.80
21	2021–22	3285	3316	54,781	25,266.512	1,421,690.40
22	2022–23	2825	2412	44,296	19,481.677	1,047,148.63
23	2023–24	1598	775	12,767	6939.236	420,766.76

Source: PMGSY site (project details year wise report)

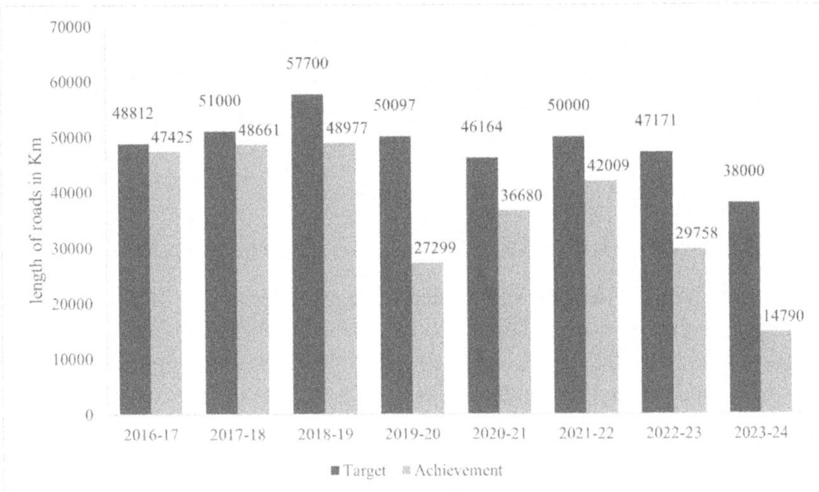

FIGURE 6.3 PMGSY Road Works

Source: Ministry of Rural Development http://rural.nic.in/netrural/rural/index.aspx

6.3 THE NEED FOR NETWORK PLANNING IN A REGION

Planning strategy depends upon several issues such as the spread of rural areas across the country, the resource base of a rural settlement, the population and growth rate of population, the distance between settlement and nearby agricultural, industrial, and administrative growth centers, and the current accessibility levels to these growth centers. Thus, public transport planning in rural areas is a complex phenomenon. It is quite difficult to consider each and every settlement in planning public transport. In India, a district is normally considered as a unit for the purpose of planning and developing transport infrastructure. The same criterion could be followed even while planning a public transport network in rural areas.

Rural public transport network development may be conveniently considered under two headings: purely rural and rural–urban transport. Purely rural transport consists of movement wholly within rural areas, to and from work and transporting goods for domestic use. Rural–urban travel is between rural locations and, usually, the nearest market town, and demand often fluctuates considerably according to the incidence of market days. The main purposes of rural–urban travel are normally business related, principally taking goods to and from the market. At this level, many journeys are a combination of passenger travel with freight transport. Thus, much of the traffic could be classified as accompanied freight.

Public transport is widely regarded as a critical infrastructure for villages to integrate with other nearby economic activity centers. Much of the focus in public transport planning is on engineering and operational considerations often linked to scheduling algorithms rather than designing an overall network that is coherent to users. Well-planned public transport supported by rigorous design and planning of the network of lines and their interconnections is essential to achieving the task of efficient connectivity between rural settlements with cities in the district. Two broad network strategies which were categorized as radial and dispersed network strategies followed in urban areas could be tried even in public transport network planning in a district.

6.3.1 Transport for School Children and College Students

Transport demand for schoolchildren and college students is a major issue in network development. As the population increases, the number of children of school age increases both in absolute terms and as a proportion of the total population. Improvements in educational facilities and midday meal programs in India to decrease school dropouts has achieved a substantial number of increases in the number of children attending school. This has resulted in the number of children traveling to and from school increasing at a much faster rate than population growth.

Distances up to ten kilometers are relatively common, while some children travel considerably farther, because of inadequate good colleges. Walking to and from the school becomes practically nil as the distance increases. Through the provision of subsidized transport many states in India are able to reduce school dropouts considerably. The provision of public transport to the requirement of the education sector is creating several difficulties to state-owned public transport operators. The demands for exclusive school trips from a settlement are low and the timings mostly differ from other economic activity requirements. This has resulted in huge losses for the operators running exclusive buses for the requirements of the education sector with subsidized fares.

Sometimes the concession is fully funded by the government, but, in most cases, the transport operators are expected to bear part of the cost themselves. Very often, transport operators, particularly those in the private sector, are unwilling to carry children at peak periods at concessionary fares, as this will mean turning away adult passengers who would pay the full fare. Public bus operators are forced to accept this subsidized travel demand. Thus, there exists a complexity in providing public transport in rural areas in many developing countries.

6.3.2 Dedicated Public Transport for Students

Many states in India operate dedicated passenger transport services for the exclusive use of their students. While these cannot be regarded as public transport services, they are relevant in the sense that a large number of students use public transport and may constitute a significant element of the passengers of transport system. Dedicated school buses may be owned by the government, educational authorities, or even by individual schools, or may be chartered by transport operators. The buses which carry students in the early morning could be used to provide other services such that the vehicle utilization and revenue earning could be maximized. However, this needs proper planning and scheduling of buses to achieve the satisfaction of many users.

6.3.3 Market Segmentation

Public transport users, even in villages, comprise a large cross-section of the public with different requirements and interested in using buses for different purposes. If only one standard of service is provided, it will meet the requirements of only one section of the population leaving perhaps significant unsatisfied demand. There is scope for more than one standard of service to cater to passengers with different requirements. Substantial requirements of the people can be covered from such segmentation of the market. Different services may have to be planned to cater to the needs of each segment. Many passengers are satisfied with the basic LOS usually provided and are prepared to accept overcrowded, uncomfortable, and unreliable vehicles, as they could not afford to pay higher fares for a better quality of services. There may be other passengers who are prepared to pay higher fares for a convenient, comfortable, reliable service, with a better quality of service. There may in fact be scope for several LOSs, ranging from those using the most basic type of vehicles, with a high passenger capacity and minimal comfort, to premium quality services offering a high degree of comfort.

6.3.4 Mobility and Accessibility

The difference between mobility and accessibility is fundamental to the understanding of rural transport provision. In the past, the thrust of policy formulation has generally improved mobility with little consideration given to accessibility issues. The modern approach to transport policy is the reverse.

Hillman et al. (1973) defines mobility as the capacity that a person has for getting around. It is clear from this definition that individual circumstances are very important. Thus, mobility depends on personal factors such as age group, purpose of travel, status in society, and financial parameters of the trip maker and upon the range of transport facilities that are available (Tolley and Turton, 1995). Mobility will change throughout a person's lifetime (typically being fairly poor during childhood, increasing in adulthood with quite likely the ability to afford a car, and then reducing once again in later stages of life). It is usually measured in terms of the number of trips or journeys made per unit of time (normally in a day or a week).

The demand for mobility and variation in mobility levels over a period of time could be obtained through primary data from different households in a village. Table 6.3 provides a matrix of mobility measurements for different purposes at different time periods.

Most of the trips in rural areas could be classified with six different purposes. The changes in the quantum of travel and the average distance for each purpose provide information regarding the economic development of the village and provision of infrastructure at or near the village.

6.3.4.1 Defining Accessibility

Accessibility is difficult to define. As Gould (1969) says, accessibility is a slippery notion, one of those common terms which everyone uses until faced with the problem of defining and measuring it. Daly (1975) defined accessibility as the ease with

TABLE 6.3
Journey per Person per Year and Average Length by Journey Purpose

	Journey per Person per Year			Average Journey Length (Miles)		
	Year 1	Year 2	Year 3	Year 1	Year 2	Year 3
Commuting for work	X11	X12	X13	D11	D12	D13
Business	X21	X22	X23	D21	D22	D23
Education	X31	X32	X33	D31	D32	D33
Shopping	X41	X42	X43	D41	D42	D43
Social	X51	X52	X53	D51	D52	D53
Recreation	X61	X62	X63	D61	D62	D63
Total	Xt1	Xt2	Xt3	Dt1	Dt2	Dt3

which people can reach distant but necessary services, while Mitchell and Town (1976) define accessibility as the ability of people to reach destinations at which they can carry out a given activity. Buchan (1992) defines accessibility as being one of the quality-of-life objectives for transport and describes the objective thus: to encourage and provide a transport system which will give people access to workplaces, to shops and public buildings, to industry and commerce, to facilities of health and other services, and to one another.

6.3.4.2 Relationship Between Mobility and Accessibility

Mobility and accessibility have been presented thus far as two completely separate concepts. In reality, however, they are interlinked. Some people in rural areas suffer from both lack of mobility and accessibility. Many people in rural areas have no access to a car and often public transport alternatives are inadequate. There are also many low-income people in rural areas who cannot afford to travel far. This is the mobility problem. For these people, however, accessibility is more of a problem because access to many facilities requires the use of mechanized transport. Thus, as far as rural development is concerned, both aspects need to be addressed if rural areas are to be living and thriving areas with economic activity. Improving mobility and accessibility for those without access to a mechanized mode in rural areas is much more difficult because of the dispersed nature of the centers of population and the fact that the solutions to the problem vary between groups of people living in given rural settlements.

In India, the problem of road accessibility is being solved on a war footing through the introduction of PMGSY. PMGSY was introduced in 2000 to ensure AWR connectivity for rural habitations. Initially, it targeted villages with over 500 residents in plains and over 250 residents in hilly or challenging terrain. In 2013, PMGSY-II expanded this goal to upgrade 50,000 km of existing rural roads. By 2019, PMGSY-III aimed to improve and consolidate 125,000 km of routes connecting villages to key social and economic facilities. An additional 10,000 km of roads were also approved to support remote and underdeveloped areas.

Overall, about 782,850 km of roads have been sanctioned since the program's start, with an investment of around Rs. 344,500 crore (US$ 45 billion). Of these, 693,000 km have been completed at a cost of Rs. 270,000 crore (US$ 36 billion), resulting in all-weather connectivity for approximately 11,561 previously unconnected habitations.

6.3.4.3 Measuring Accessibility

Accessibility is usually measured in terms of journey time or cost, or a composite of the two (sometimes referred to as generalized cost). Thus, accessibility to a hospital can be measured in terms of how long it takes to get there or how much it costs to get there, or a mixture of the two. For some people, the time taken is the most determining factor, for others (pensioners for example), it may be the cost which is most critical in making the journey.

In India, as a high proportion of rural areas (65%) reported having a bus stop and a grocery shop within a relatively short walking distance (6 minutes), accessibility could be considered as satisfactory. The quantification of accessibility to bus services in rural settlements with populations between 500 and 2000 based on the LOS needs to be properly understood and planned. The scoring system is outlined below with evening and Sunday scores added together. The higher the numerical value in given in Table 6.4 indicates higher satisfaction.

6.3.4.4 Perspectives on Mobility and Accessibility in Policy

As discussed elsewhere, the rural transport policy needs to consider both the people who live in rural areas and those who visit areas for economic activity. The needs of the two groups are likely to be quite different and this needs to be reflected in policy development. The first step is to review the rural transport problem. This is illustrated in Figure 6.4. The policy required could be depicted diagrammatically as shown in Figure 6.5.

TABLE 6.4
Quantification of Satisfaction for Planning Rural Public Services

Daytime Services		Evening and Sunday Services	
No service	0	None	0
One day a week	1	1 return journey before 6 pm	1
Peak service, 6 days a week	2	As above plus 2 Sunday returns	2
Peak service and shopping, 6 days a week	3	Better than 2 hours in the evening and	3
Four return journeys per day	4	Sunday services	
Hourly or better service	5		
30 minutes or better	6		
20 minutes or better	7		

Source: TAS (1997)

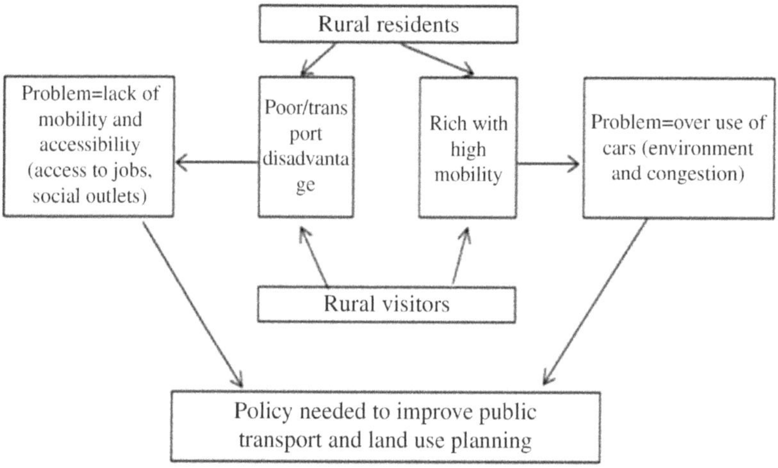

FIGURE 6.4 Rural Transport Problem

Source: Stokes and Cullinane, 1998, Rural Transport Policy

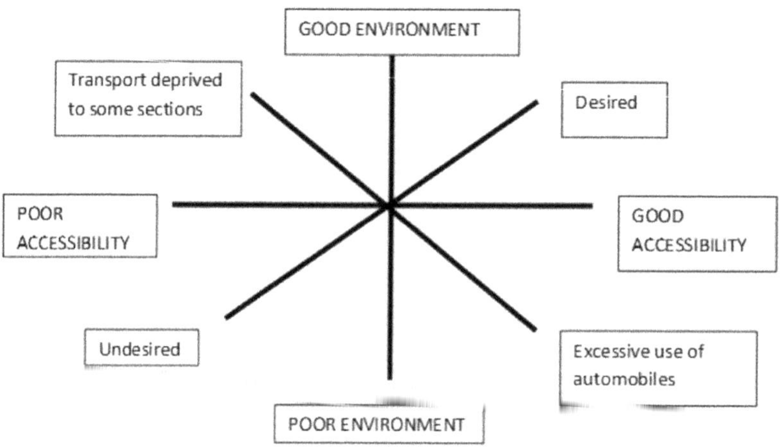

FIGURE 6.5 Rural Transport Policy Requirements

Source: Stokes and Cullinane, 1998, Rural Transport Policy

The interrelation between the desired mobility and desired accessibility could be better explained through Figure 6.6. It is often assumed that mobility and accessibility refer to the same thing, that is, that mobility equals accessibility. This is shown by the 45-degree line in Figure 6.6. However, policy should be aimed at reducing the usage of automobile traffic but increasing accessibility to that area.

However, in order to reduce environmental degradation, the mobility levels need to be planned properly. Currently, the environmental problem does not exist in rural

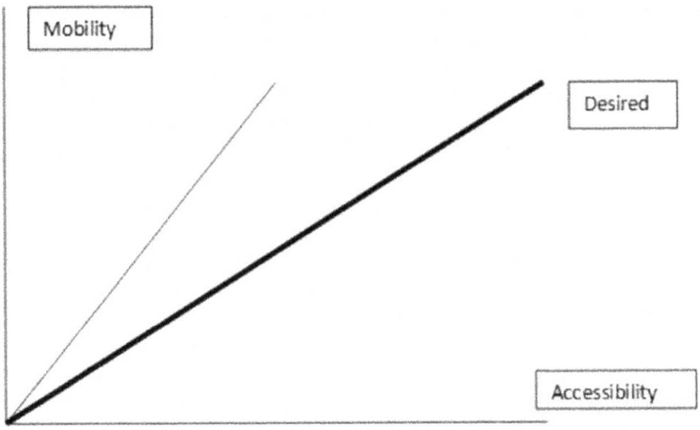

FIGURE 6.6 Relationship Between Mobility and Accessibility

Source: Stokes and Cullinane, 1998, Rural Transport Policy

areas or in cities with less than 0.5 million population. However, if steps are not initiated to promote public transport vigorously, proliferation of mechanized modes for intradistrict and interdistrict travel may create environmental problems in the next two decades even in rural India. The relation between mobility and environment could be diagrammatically presented as shown in Figure 6.7.

With a given technology, environmental problems increase with increasing mobility. Policy should be targeted at reducing the environmental impact of any given level of mobility. This could be done, for instance, by improving technology or by increasing the use of environmentally less damaging vehicles such as buses or trains. The desired policy option is again depicted by the solid line.

There is a relation between the income levels of different sections of society and the accessibility levels they enjoy. This relation is diagrammatically represented in Figure 6.8.

Accessibility increases with income as richer people have access to better and improved transport technology, and they can afford to pay for services. Policy should be aimed at increasing accessibility for all levels of income groups. All sections of the public should enjoy a basic level of accessibility and mobility. This could be achieved through increased provision of public transport. The desired policy line is again depicted with a thick solid line in Figure 6.8. This policy helps in the development of an inclusive society.

6.3.5 Benefits of Public Transport Network Development

The various benefits of implementing an effective rural public transport network are presented in Table 6.5.

Thus, in order to bring the rural areas into mainstream of economic activity, there is an urgent need that all the rural areas in the country be connected with reliable and sustainable public transport network. In order to achieve this sustainable mobility between rural areas with other nearby towns, a national policy may be

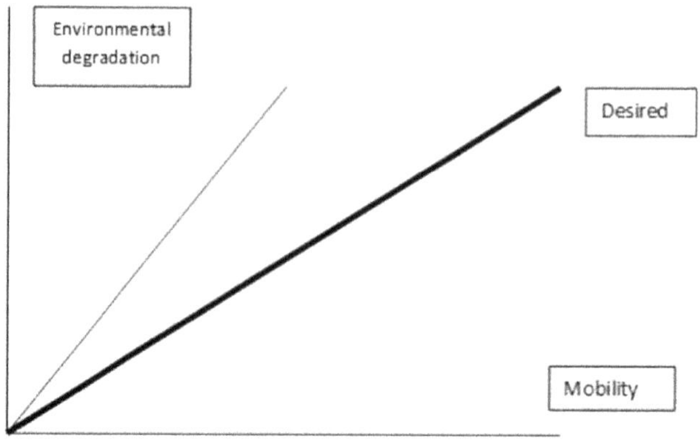

FIGURE 6.7 Relationship Between Mobility and the Environment

Source: Stokes and Cullinane (1998), Rural Transport Policy

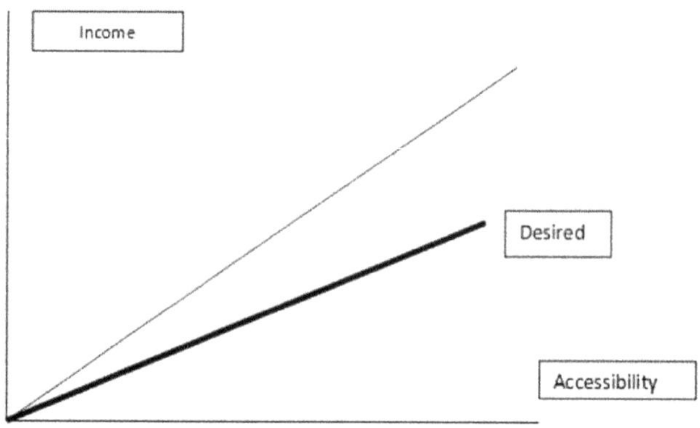

FIGURE 6.8 Relationship Between Accessibility and Income

Source: Stokes and Cullinane, 1998, Rural Transport Policy

more appropriate. The guidelines are needed to develop integrated public transport networks to follow certain criteria, norms, etc. This national strategy will provide guidelines for different states in its implementation. The policy should aim at the following aspects:

1. Public transport strategy and an action plan for its implementation during the next 10–15 years.
2. Development of capacity building in village panchayats as well as in municipalities.
3. Identification of the areas of subsidy and quantum of subsidy required to provide sustainable integrated public transport network planning in rural areas.

TABLE 6.5
Benefits of Rural Public Transport

Item	Benefit
What is the impact of public transport impacts of the proposal?	Public transport will benefit many rural households not owning vehicles.
What will be the impact on the environment?	The combination of modal switch to public transport will lead to a reduction in pollution. The investment package will target environmental improvements within the district.
What are the impacts in terms of transport economic efficiency?	There is a net economic benefit for all when public transport improvements are added.
What will be the local impacts in terms of employment?	Rural areas are likely to develop economically.
What will be the impact on accessibility?	The investment package will improve accessibility to/from the rural villages.
What will be the impact in integrating transport modes and services?	Integration of transport modes and services will benefit the investment climate of rural settlements.
What will be the impact of the proposal against wider government policy?	The proposals integrate with wider government policy on health, education, and the environment.

4. Wherever this is partly in existence, the delivery gaps need to be identified and readdressed.
5. This integrated public transport network may be coordinated with NMT infrastructure and other transport activities.
6. Piecemeal implementation with a large number of gaps may compromise the effectiveness of the system.

6.3.6 Strategy Followed in South Africa in the Development of Rural Public Transport Network

In South Africa, for the benefit of rural areas, an integrated public transport network (IPTN) is put in place under a national guiding framework (NGF). This NGF identifies and focuses on planning and implementation of rural transport network.

1. It will address the implementation gaps and overlaps and consolidate resources and funding.
2. It will address access and mobility through an integrated public transport management model.
3. It will document the best local practices and provide for appropriate interventions.
4. It integrates transport infrastructure and services through network management.

After its implementation, the improved provision of public transport has achieved the following outcomes.

1. Key Performance Indicators (KPIs):
   Acceptable standard of service
   Reduced travel time to work/education and general trips
   Acceptable walking time to public transport service
   Acceptable waiting time for public transport service. Acceptable cost of transportation.
2. Enhanced social and economic activities:
   The improvement of transport systems will have multiplier effects of increasing accessibility, reducing poverty, and enhancing social and economic development.
3. Improved connectivity with good provision of infrastructure:
   Safety and reliability of public transport services, differentiated choice of modes, organized and licensed operations, safe roads, and reduced level of accidents
4. Improved status of rural transport planning in the district, responds to new demands.
5. Aligned and well-coordinated transportation services/operations, well-defined roles for all transport players
6. Reduced level of migration to the metropolitan municipalities, social stability, and cohesion
7. Rejuvenating the local economy and tax base and sustaining rural livelihood:

The above example in a developing country has achieved a number of benefits. Thus, it is more appropriate to follow a similar strategy in India.

6.3.7 Strategy for Rural Transport Network Development in India

Rural settlements are normally spread across the entire nation. India is no exception to this phenomenon. In order to develop a comprehensive rural public transport network, evolving a national policy may be the first step. In India, PMGSY policy addresses rural road connectivity which is already in progress. Public transport network development invariably requires good AWR networks, it is appropriate to consider rural network as a starting point to plan public transport network covering rural settlements. Most of the states in India are already involved in providing public transport connectivity to many villages. However, what is really needed to study is whether this fits into integrated and sustainable rural development. In India, while implementing PMGSY, a core network for a district is prepared connecting all the settlements. This may be considered as a first step in planning public transport network in the district. If required, additional bus routes from a particular village may be initiated. A comprehensive rural public transport network development may be initiated at the national level through legislation in the parliament, similar to that of PMGSY. This policy helps in providing resources for implementing the national strategy.

The benefits of this strategy could be summarized below:

While developing core public transport network for the entire district covering all settlements, the following criteria may be followed.

1. *Simple and Direct Network Structure:* Every village as per its requirements should be connected to nearby economic activity centers.
2. *Plan a Hierarchy of Routes into a Network:* As the demand from rural areas will be meager, a hierarchy of routes may be developed so that from a given settlement to reach far-off places, a transport at a nearby economic center will provide opportunities for long distance travel.
3. *Plan for Consistency and Reliability:* This will ensure that rural households can develop upon this network plan with confidence and may increase the patronage in the long run.
4. *Coordinate Convenient Transfers:* This policy will help to minimize long waiting times for multiple journeys.
5. *Integrate Fare Systems with the Full Journey:*
   Normally long-distance travel by a single bus is less costly than journeys with multiple boardings at two locations. In order to encourage more and more people to use a newly initiated public transport network the fare structure may be integrated. This will help in the long run to achieve better occupancy ratios and maximize income to public transport operators. This may be attempted initially at least to move to neighboring districts if not state headquarters.

6.4 TRAVEL DEMAND ESTIMATION IN A REGION, SAY DISTRICT

There are two terms associated with the measurement of mobility of family or an individual. These two terms are NEED and DEMAND. NEED is a social concept, and DEMAND is an economic concept. This distinction could be better understood by considering the food intake of different sections of society. For healthy growth, every individual requires to take a minimum quantity of minerals, carbohydrates, and other nutrients. This criterion provides the true estimation of the NEED. However, what a family consumes on an average per day provides the DEMAND for food items of that family. If you compare this DEMAND with NEED for different sections of people, there will be wide fluctuations. In some cases, the DEMAND is much higher than NEED and in some other sections the DEMAND is much below the NEED. In a society with inclusive growth objectives, the difference between DEMAND and NEED should be nil or positive. If you apply the same concept to transport-related issues, every individual has to undertake some minimum number of trips to achieve social and economic objectives/goals. However, it is very difficult to measure the NEEDS for travel of different sections of society. One simple way to measure this is to make public transport free to all individuals. In the long run, the demand for travel under such a situation can be taken as NEED for travel. The actual quantum of travel at the current prices provides DEMAND for travel. This process also provides answers to the impact of price on DEMAND.

In addition to this philosophical distinction, the actual quantum of travel by public bus depends on routing and scheduling of public bus services from that settlement with that of surrounding villages and towns. Bus transport operators, both public and private, need to consider this aspect while estimating the fleet strength.

6.4.1 Classification of Transport Services

The passenger road transport services in general could be broadly divided into four categories on the basis of different characteristics as follows:

1. Interstate
2. Interdistrict
3. Intradistrict and rural
4. Urban operations

The analysis in this section broadly follows this classification as the characteristics of these segments vary requiring different interventions. The characteristics of each of these services vary widely with respect to vehicle utilization, time of operation, crew requirements, profitability, and major players for the types of services. This categorization means that these categories require different types of policy and financial support (as shown in Table 6.6).

The demand is analyzed for the various sectors of urban operations, intradistrict, rural services, interdistrict services, and interstate operations. The characteristics of each one of these services are quite different with respect to frequency of service, time of service, service quality, willingness to pay, and the expectations of the public, etc.

TABLE 6.6
Classification of Transport Services According to Financial Support and Different Types of Policy

Category	VU (km/day)	Service Types	Players	Profitability	Main Competitors to STUs
Interstate	>600	Volvo, luxury deluxe	Public and private	Very high	Private
Interdistrict	300–450	Luxury, express, Volvo	Public and private	Moderate and high	Private
Intradistrict	250–350	Express, ordinary, limited halt	Mostly private, maxi cabs	Low and loss making	Illegal contract buses, maxi cabs
Urban	200–250	Ordinary, limited halt	Mostly private autos	Mostly loss making	Illegal contract buses, maxi cabs

6.4.2 Demands for Intradistrict and Rural Services

In any road transport undertaking, the maximum number of buses operate as intradistrict and rural services. These rural operations provide two major categories of services. One, by providing accessibility to villages with administrative and economic activity centers at the district level. Two, by providing service within the district between and among the places of importance such as business locations, historical places, linking taluk headquarters and district headquarters, and interlinking other major economic activity locations. Thus, the demand for these services is linked to:

1. Number of settlements in the state
2. Number of economic activity centers in the state

However, it is difficult to estimate the number of activity locations in the state as their number changes and grows with the development of the state. The proxy to this is the size of the rural population undertaking journeys in the state. The number of trips in rural settlements will be much less compared to urban areas. One bus on average will be able to cater to four rural settlements. In addition, it is felt that 60 buses may be required per million of the population to link various economic activity centers in a state.

These norms have been applied to the state of Andhra Pradesh to validate the model in a study done by IIMB. The demand for services as per the model suggested works out to about 10,250 buses, given the number of settlements at 26,613 and estimated rural population of 60.1 million as of 2006. However, APSRTC ran about 10,000 buses to cater to this segment. The survey conducted by IIMB in nine districts of Andhra Pradesh clearly proved that the services are adequate in the sample districts. In addition, a small number of private operators are also providing service in this segment in the state of Andhra Pradesh. This norm is also applied to estimate the fleet strength in three public bus corporations of Karnataka State and the results are in tune with the opinion of policy makers and officials of KSRTC. If this norm is applied at all India levels, the demand for bus fleet to cater to this demand during different time periods is presented in Table 6.7. Currently, the deficit in this segment is quite high to the extent of 50% of the requirement.

6.5 STAKEHOLDER CONSULTATION AND PREFERENCES TO OPTIMIZE PMT ROUTES

Public bus transportation plays a pivotal role in India in bringing about greater mobility both within and between rural and urban areas. Through increased mobility, it also contributes immensely to social and economic development of different regions of the country. Realizing the importance of this link, the Government of India, in course of time, has invested heavily in the development of a network of bus services to link up towns and villages all over the country.

The 70 public STUs in India today own more than 150,000 buses. These undertakings together operate 17,000 million passenger-km daily and serve 70 million

TABLE 6.7
Bus Fleet for Intra and Rural Operations in India

Item	2006	2011	2016	2021
Number of Settlements	600,000	600,000	600,000	600,000
Buses at 1 for 4 Settlements	150,000	150,000	150,000	150,000
Rural Population (million)	805	874	950	923
Buses at 60 per million	48,327	52,484	56,998	55,715
Total Bus Fleet Needed	198,327	202,484	206,998	205,715
Share of SRTU at 50% Level	99,168	101,242	103,499	102,858

passengers a day. The STUs in the country are set up under four forms: (1) departmental undertaking directly under the state governments, (2) municipal undertakings owned and controlled by the municipal corporations, (3) companies formed under the Indian Companies Act 1956, and (4) road transport corporations formed under the RTC Act 1950. While the objectives that the STUs set out to achieve are laudable and, indeed, the effort began in right earnest, the support needed from state and central governments to sustain the effort suffered a setback beginning in the 1970s. Currently, while the need of STUs to meet its social obligation is retained, the budgetary support to achieve the objective has drastically shrunk over time. There is a necessity to obtain an insight into the needs and expectations of the public in respect of passenger transport in general and rural areas in particular. In addition, the current level of customer satisfaction with the transport services provided can play a significant role in network planning of public transport services in rural areas. Network planning involves two major components, namely identification of route and frequency of operations on the route. Before the development of public transport network connecting various settlements, the operators need to gauge the perception of commuters to balance the need/demand for travel and the resources available with the operator. Stakeholders of public transport judge the operators on the basis of the quality of service enjoyed by them.

Quality is an abstract entity and subjective in character. It is simply an expression of the extent to which the service provided by an operator fulfills the expectations of the customers/stakeholders concerned. As such, there will be variations in perception of quality of service enjoyed by different customers for the same operation. The term "quality" may include many components such as:

1. Comfort and convenience
2. Schedule and operations
3. Crew behavior
4. Cost and other factors

Each of these components could include many attributes and the same are included in Table 6.8.

TABLE 6.8
Attributes Considered for Service Quality

Comfort and Convenience	Schedule and Operations	Crew Behavior	Cost and Other Aspects
Overloading	Notification of schedules	Courteousness with passengers	Notification of fares
Boarding and alighting	Following the schedule	Helping children and old age people	Returning small changes
Seating arrangement	Prompt service during breakdown	Rash and negligent	Adequacy of fares
Movement within the bus	Maintenance of vehicles	Appearance of the crew	Charges for luggage
Driving comfort	Cancellation of schedules	Neatness and professionalism	
Travel time	Arrival/departure timings	Attitude of the crew in general	
Luggage allowance			
Stopping at the bus stops			

In order to gauge the perceptions, primary data needs to be collected from the stakeholders on the attributes listed above. Unfortunately, there is no evidence that this exercise is undertaken periodically by any public or private operator. IIMB was involved in one such exercise for Karnataka State Road Transport Corporation (KSRTC) along with several bus operations provided by private operators in the state of Karnataka. The results indicate that the satisfaction levels on most of the attributes are much above for KSRTC compared to any private operator. IIMB was also involved in obtaining the quality of service provided by APSRTC. The results indicate that commuters are quite satisfied with the operations of APSRTC.

6.6 IMPACT OF TRANSPORT INFRASTRUCTURE IN THE DEVELOPMENT OF RURAL INDIA—A CASE STUDY

Many research studies and administrative reports emphasized the need for better transport infrastructure for accelerated economic development of India. Both urban India as well as rural India is experiencing inadequacies of transport infrastructure.

Urban areas mainly suffer from a shortage of transport networks to meet the ever-increasing demand. This is reflected in *v*/*c* ratio exceedingly more than one in most of the urban corridors in India. This problem is much more severe in almost all cities with a population of more than 5 million. A recent study by IIMB revealed that in all the eight important corridors of Bengaluru city the *v*/*c* ratios are more than one. The impact of these high *v*/*c* ratios are reflected in terms of congestion, delay, accidents, and economic losses. Currently, the Ministry of Urban Development

(MOUD) is involved in improving the urban mobility through dedicated funding under JNNURM to selected cities.

In respect of rural areas, the problem is quite the opposite, that is, many settlements do not enjoy AWR connectivity. In India, we use the term AWR to provide accessibility throughout the year including monsoon season. If a settlement is accessible during part of a year, especially during non-monsoon period, then it is considered to have been accessible by fair-weather roads (FWR).

During the decade 1975–85, the Ministry of Transport, Ministry of Rural Development, and the Planning Commission initiated several rural road studies in a number of districts to assess the socioeconomic scenario of FWR and AWR connectivity patterns. IIMB has entrusted the Ratnagiri district in Maharashtra to quantify the socioeconomic impacts of AWR and FWR in that district. Ratnagiri is a land-locked district in the state of Maharashtra. As per census information, 98% of settlements were provided with AWR systems. However, the geographical distribution of houses in a village in Ratnagiri district is quite different from other districts of India. A normal village consists of six to eight wadies (wadi is part of a village) separated by 3–5 km between wadies.

In the census, it is normally recorded as having an AWR if the main wadi is connected by an AWR. However, all the wadis might not have been really enjoying an AWR. The comprehensive study undertaken by IIMB revealed the following patterns:

1. The average income of a household in a wadi (part of a village) connected by an AWR is statically higher than the household which enjoys FWR accessibility.
2. At the taluk (part of a district with a number of settlements) level also, this hypothesis was noticed and statistically proved.
3. This study was repeated in other districts of Maharashtra. Even at the district level, districts with a higher percentage of settlements connected by AWRs, enjoy higher per capita income.

Thus the study clearly brought out that settlements connected to AWRs enjoy statistically higher income at the macro level, that is, district level; sub macro level, that is, at taluk level; at micro level, that is, at settlement level; as well as sub micro level, that is, household level in a settlement.

IIMB was also involved in another rural road study sponsored by the Ministry of Road Transport and the Ministry of Rural Development across southern part of India covering eight states. The study was conducted during three time periods: first time in 1976–78, second time in 1986–87, and finally third time during 1996–97. Many important findings were observed. For proper comparison, in all three studies the same sample frame was (the same settlements) used.

The sample could be broadly divided into the following:

1. Villages remain as they were in the first study until the end of the third study, that is, FWR as FWR and AWR as AWR.
2. Change of accessibility from FWR to AWR.

3. Change of accessibility from AWR to improved connectivity to class 1 roads during this 20-year period, that is, major district roads or state highway system.
4. Settlements with multiple connections to class 1 category of roads.

The results of all these studies indicate that during the 20-year period (1976–96) locations with category 1 mentioned above saw a quantum of change in both passenger demand and freight demand at the least. The changes are more pronounced (quantum jump) in the case of category 2. Exponential growth was recorded in categories 3 and 4.

The freight movement pattern also exhibited many major changes. There is a gradual shift from bullock carts (which account for more than 75% of the total tonnage) NMT movement during 1976 study to mechanized models in 1996 (accounting more than 80% of the total tonnage). The shift also is more pronounced in the settlements with improved connectivity patterns.

Many international agencies are also providing funds to improve the rural accessibility in India. The state of Andhra Pradesh received assistance from the World Bank under the Andhra Pradesh economic reconstruction project in three districts of Andhra Pradesh. At the end of the project, World Bank insisted that an impact study needed to be undertaken by a third agency. IIM was again roped in to undertake this task.

The main objectives of the study were:

1. To assess the direct and indirect socioeconomic impacts/benefits of the rural roads.
2. To study the impact of these new roads on the travel characteristics of the rural population.
3. To carry out an economic analysis comparing the costs and benefits of these roads.
4. To assess whether the expectations of the village communities and local users are met.

The sample details for the study are presented in Tables 6.9 and 6.10.

The responses reveal that there are considerable improvements in many areas of socioeconomic indicators as presented in Table 6.11.

The results clearly show that all sections of the household are satisfied. However, there are variations among districts, road type, and settlement size.

The results clearly indicate that education, health, and banking sectors are positively impacted by the changes in accessibility pattern. Other benefits include:

1. Travel cost savings due to laying roads
2. Travel time saved per day by the road users
3. For calculating monetary value of travel time savings, 25 working days in a month and Rs. 90/day wage are used. Thus, an hourly rate of Rs. 9 is used.
4. The average monthly travel cost savings on average is Rs. 127.50, and travel time saved is about 22.61 minutes.

TABLE 6.9
Summary of Sample Data Collected

Description	Number
Total no. of districts represented	3
Total no. of villages considered	37
Total road users	903
Total households	620
Total banks	6
Total schools	32
Total healthcare centers	20

Source: censusindia.gov.in

TABLE 6.10
Distribution of Sample Villages by Population

Population Ranges	Warangal	Karimnagar	Adilabad	Total	Percentage of Villages Selected
<1000	0	0	2	2	5.4
1001–1500	3	4	2	9	24.3
1501–2000	1	2	2	5	13.5
>2000	9	6	6	21	56.8
Total	13	12	12	37	100

Source: censusindia.gov.in

TABLE 6.11
Number of Road Users Before and After Laying Roads by Mode of Transport

No.	Mode of Transport	Before	After
1	Pedestrian	399	123
2	Public transport	53	133
3	Private transport	6	20
4	Auto/taxi	82	174
5	Freight transport	13	31
6	Bullock cart	13	18
7	Cycle	184	195
8	2-Wheeler	43	202
9	4-Wheeler	1	7
10	Others	54	0
11	Boat	1	0
12	Not specified	54	0
	Total	903	903

The distribution of sample trips by mode before and after improvements is presented in Table 6.12.

The benefit stream consists of three components: Benefits accrued to cultivators for better yields and prices as informed by them, savings to road users in their travel costs, and travel time saved converted into monetary values. This travel time savings was considered in two ways. In Case A, whatever travel time is saved, only 10% of it could be converted into earning additional wages. Thus, an individual who earns ten hours in a month may be able to earn additional wages for one extra hour only in a month. In Case B, it is possible to utilize the savings fully in realizing additional wages. The costs and benefits for the study villages as a whole are presented in Tables 6.13 and 6.14.

The analysis clearly shows that for Case A, the Economic Internal Rate of Return (EIRR) is about 18.71% and for Case B it is about 39.39%. It may be noted from the figures that the payback period is just three years for Case B and six years for Case A.

TABLE 6.12
Benefits Recorded (%) by Settlement Size

S. No.	Benefits	All	Population of Villages			
1	Women	All	501–1000	1001–1500	1501–2000	>2000
	Transport facilities are good and increased	42.41	25	51.79	13.33	47.54
	Ease of travel between villages	9.38	0	10.71	6.67	10.66
	Good for women to travel during maternity period	17.86	31.25	10.71	36.67	14.75
2	Children					
	Good/increase in transport facilities	26.11	20	17.07	13.04	33.02
	Good facilities for school children	51.11	60	56.1	47.83	49.06
3	Aged					
	Good/increase in transport facilities	29.17	22.22	31.71	4.35	34.74
	Convenient to go to hospital	35.12	33.33	31.71	52.17	32.63
4	Students					
	Good for higher education/further studies	26.35	28.57	27.5	27.78	25.3
	Easy transport facilities for college going students	43.24	28.57	37.5	44.44	46.99
5	Sick persons					
	Good for emergency service/treatment	22.4	20	18.75	15.79	26.09
	Convenient to go to hospital	48.8	60	43.75	52.63	49.28

TABLE 6.13
Cost–Benefit (Case A)—All Districts (Rs. Million)

Year	Cost	Benefit	Net Benefit
0	360	65.75	–294.25
1	370.8	132.82	56.27
2	381.92	201.22	57.28
3	393.38	271	58.32
4	405.18	342.17	59.37
5	417.34	414.76	60.44
6	429.86	489.17	61.89
7	442.75	565.06	63
8	484.84	642.48	35.33
9	498.52	721.44	65.28
10	512.61	801.98	66.45
11	527.12	884.94	68.44
12	542.07	970.39	70.5
13	557.47	1058.4	72.61
14	573.33	1149.05	74.79
15	589.67	1242.42	77.03
Internal Rate of Return:			18.71%

TABLE 6.14
Cost–Benefit (Case B)—All Districts (Rs. Million)

Year	Cost	Benefit	Net Benefit
0	360	105.55	–254.45
1	370.8	213.21	96.86
2	381.92	323.03	98.69
3	393.38	435.04	100.55
4	405.18	549.29	102.45
5	417.34	665.82	104.38
6	429.86	785.27	106.93
7	442.75	907.11	108.94
8	484.84	1031.38	82.19
9	498.52	1158.14	113.08
10	512.61	1287.44	115.2
11	527.12	1420.61	118.66
12	542.07	1557.78	122.22
13	557.47	1699.07	125.89
14	573.33	1844.59	129.66
15	589.67	1994.48	133.55
Internal Rate of Return:			39.39%

SUMMARY

This chapter highlights information about public transport planning in rural areas, which mainly covers topics like planning strategy in which geographical settings of rural areas are discussed and also the information consists of network planning for school and college students and also it deals with the market segmentation, the relation between mobility and accessibility.

7 Economics of Public Mass Transport

7.1 ENERGY EFFICIENCY OF PMT WITH RESPECT TO PERCENTAGE CONTRIBUTION OF PMT

Transportation is central to addressing many economic and social development challenges, yet it remains a significant contributor to global energy consumption and GHG emissions. Currently, the sector accounts for about 64% of global oil consumption, 27% of total energy use, and 23% of the world's energy-related carbon dioxide emissions. With transportation responsible for over one-fifth of global GHG emissions, transitioning from private cars to public mass transit presents a promising pathway for mitigating environmental impacts. However, for this shift to be effective, public transit must be a convenient and accessible option for travelers.

Public mass transit systems—such as buses, trains, metro systems, and shuttle services—offer shared mobility solutions that emit far fewer climate-warming GHGs per passenger compared to private vehicles. For instance, while cars emit between 57 and 322 grams of carbon dioxide-equivalent per passenger-km $(gCO_2\text{-eq/pkm})$, buses emit just 22–92 gCO_2-eq/pkm, and trains emit 6–118 gCO_2-eq/pkm, according to the International Energy Agency. The efficiency of these systems lies in their ability to transport large numbers of people simultaneously, significantly reducing the per capita energy consumption and emissions compared to individual car travel.

Despite its potential benefits, public transit often faces criticism for being an inefficient means of reducing fuel use and emissions. Detractors argue that the average fuel consumption per passenger-mile for public transit is only modestly lower than for cars and may even exceed that of highly efficient vehicles like hybrids. They contend that promoting the adoption of fuel-efficient and alternative-fueled vehicles could be a more cost-effective strategy.

However, Litman (2012) highlights how critics often fail to consider the context and objectives of different transit services. Public transit plays dual and sometimes conflicting roles within transportation systems:

1. *Basic Mobility:* This involves providing services that are widely distributed, covering areas with low demand and operating during off-peak times. It also includes specialized mobility options like demand-response buses, which cater to individuals without access to private vehicles.
2. *Efficient Urban Transport:* This focuses on high-demand routes and peak travel times, offering an efficient alternative to private cars to alleviate congestion, reduce pollution, and minimize energy use.

DOI: 10.1201/ 9781003740247-7

Criticizing basic mobility services for low energy efficiency overlooks their primary goal: ensuring accessibility for all, regardless of demand patterns. Similarly, criticizing urban transit for favoring higher-income passengers neglects the necessity of offering high-quality services to attract travelers who might otherwise drive. Author further highlights how efficiency in public transit is highly dependent on load factors, or the number of passengers per vehicle. A lightly occupied bus may consume more energy per passenger than a car, but a fully loaded bus carrying 50 passengers can be up to ten times more energy efficient than an average automobile. Similarly, rail transit systems, which are three times more efficient than diesel buses, demonstrate significant environmental benefits, particularly when operating at high capacities in urban areas.

The reason for this disparity lies in the fundamental efficiency of shared mobility. While private cars typically carry one or two passengers, buses and trains can accommodate dozens or even hundreds of passengers at a time. This scalability enables mass transit to achieve significantly lower energy consumption and emissions per passenger-km, especially during peak periods or in densely populated areas.

The percentage contribution of public mass transit to energy efficiency depends on its share within the broader transportation system. In cities with well-developed public transit networks, mass transit substantially reduces energy use and emissions. However, in cities with limited or inefficient transit infrastructure, private vehicle reliance remains dominant, resulting in higher energy consumption.

In many of the developing economies such as India, for example, the modal share of public transport has declined over the years, leading to increased energy-intensive private vehicle usage. To counter this trend, cities are investing in metro systems, electric buses, and BRT corridors, which can enhance energy efficiency and reduce environmental impacts. The adoption of electric buses, in particular, has the potential to lower emissions further, provided the electricity is sourced from renewable energy.

Ultimately, transitioning to energy-efficient public mass transit is essential for achieving sustainable development goals and addressing climate challenges. By prioritizing investments in high-quality transit systems and integrating them with other modes of transportation, cities can significantly reduce their energy consumption, lower GHG emissions, and create more sustainable urban environments.

Case Study—Bengaluru Metropolitan Region

Verma et al. (2018) conducted a study to develop and evaluate sustainable transport measures aimed at improving the liveability of Indian cities, including Bengaluru, Delhi, and Mumbai. This section focuses on the case study of the BMR, which encompasses Bengaluru urban district, Bengaluru rural district, and Ramnagara, covering approximately 8005 sq. km. BMR is one of the fastest urbanizing metropolitan areas, with a 584% increase in the city's built-up area over the past four decades.

The growing urban population has led to a surge in vehicle usage, overwhelming the city's limited infrastructure and resulting in traffic congestion, longer travel times, and increased pollution, which have negatively impacted the quality of life. Additionally, issues such as land concretization, encroachment on water bodies, and inadequate maintenance of drainage systems have caused higher surface runoff on roads. This has frequently brought the city's transportation system to a standstill, further reducing its resilience. This study focuses on evaluating sustainable transport mitigation and adaptation measures to improve Bengaluru's liveability by addressing traffic congestion, reducing exhaust emissions (PM, CO, NO_H, HC, etc.), cutting GHG emissions (CO_2), decreasing carbon emission intensity relative to GDP growth, increasing the consumer surplus of sustainable transport modes, and enhancing the resilience of the city's transportation system. The study's objectives include developing mitigation strategies for the transportation sector to reduce emissions and congestion under two scenarios: "Business as Usual" (BAU) and "Sustainable Transport," covering the base year and future years 2030 and 2050, respectively. Additionally, it seeks to identify climate change vulnerabilities in transportation infrastructure and assess its impacts. Adaptation strategies are developed for the same timeframes to evaluate the scope, severity, and effects of flood events on the transport sector. Ultimately, the study aims to enhance Bengaluru's overall liveability through sustainable and resilient transport solutions.

In Bengaluru, public transport dominates travel preferences, with a mode share of 49.7%, followed by two-wheelers at 28.9%, NMT at 15.4%, three-wheelers (autos) at 3.7%, and cars accounting for just 2.3%. The total VKT in BMR is approximately 31 million for the base year, projected to rise to 48 million by 2030 and 72 million by 2050—a 60% increase from the base year. To address these challenges, four policy bundles were developed based on Intergovernmental Panel on Climate Change (IPCC) definitions, stakeholder inputs, and Delphi surveys involving Bengaluru government officials. These policy bundles aim to balance push-and-pull strategies by encouraging public transit and sustainable travel modes, thereby reducing VKT, emissions, and traffic congestion while improving the city's quality of life.

Each bundle combines various policy instruments. Bundle 1 focuses on planning and regulatory measures, while Bundle 2 combines economic and regulatory instruments. Bundle 3 integrates planning, regulatory, and economic tools, and Bundle 4 takes a comprehensive approach by including planning, regulatory, economic, and technological measures, such as full electrification of cars and buses for future target years.

The policy bundles designed for mitigating transportation challenges in Bengaluru include a range of measures to promote sustainable travel, reduce emissions, and enhance overall mobility. These bundles, categorized into four groups, aim to integrate various instruments to achieve the city's sustainability goals effectively.

Bundle 1 focuses on foundational measures to encourage sustainable travel. Policies under this bundle include expanding public transit network coverage, enhancing cycling and walking infrastructure, and imposing an additional tax on vehicle purchases to discourage private vehicle use.

Bundle 2 emphasizes regulatory and economic measures aimed at reducing emissions and improving vehicle standards. Key policies include an additional tax on vehicle purchases, stricter vehicle inspection standards coupled with improved emission norms, and an increase in fuel costs to disincentivize excessive private vehicle usage.

Bundle 3 takes a comprehensive approach by combining planning, regulatory, and economic instruments. It incorporates the expansion of public transit networks, the introduction of car-restricted roads, congestion pricing mechanisms, park-and-ride facilities, and improved cycling and walking infrastructure. Additionally, this bundle promotes carpooling, high-occupancy lanes, and mixed-use high-density developments along major transport corridors to reduce travel demand and congestion.

Bundle 4 builds on Bundle 3 by integrating technological advancements, primarily focusing on electrification. Along with all policies in Bundle 3, this bundle assumes that all cars and buses will run on electricity, significantly reducing sGHG emissions and reliance on fossil fuels.

In addition to the policy bundles, four distinct energy mix scenarios are assumed for the target years to evaluate the impacts of different energy sources on Bengaluru's transportation sector. These scenarios are as follows:

Scenario 1: New Policies Scenario (IEA, 2015)—In this scenario, 74% of electricity is derived from nonrenewable sources, and 26% from renewable sources, reflecting a gradual shift toward cleaner energy.

Scenario 2: Electricity from Nonrenewable Sources (100%)—This assumes that all electricity used in the transportation sector is generated solely from nonrenewable sources, resulting in a higher environmental impact.

Scenario 3: Balanced Energy Mix (50%–50%)—Here, half of the electricity comes from renewable sources and the other half from nonrenewable sources, representing a significant move toward sustainability.

Scenario 4: Electricity from Renewable Sources (100%)—This scenario envisions a fully renewable energy supply, significantly reducing GHG emissions and environmental degradation.

The policy bundles were rigorously evaluated across various locations in BMR. Among these, Bundle 3 and Bundle 4, which incorporate a comprehensive mix of seven policies, delivered the most favorable outcomes. These bundles showed significant improvements in reducing VKT, increasing the share of public transport and NMT, and cutting emissions.

The greatest reduction in emissions was observed with Bundle 4 under Scenario 4, where electricity is assumed to be entirely generated from renewable sources. This scenario achieved remarkable results, including nearly a 98% reduction in CO_2 emissions for both 2030 and 2050 compared to their respective BAU scenarios.

Figures reveal a consistent trend across all policy bundles: private transport mode share decreases, while public transport and NMT mode shares increase. Bundle 3 and Bundle 4 outperform others in these metrics. The rise in public transport mode share, coupled with higher occupancy levels, leads to a substantial reduction in total VKT compared to the BAU scenario for 2030. As emissions are closely tied to VKT, a noticeable decrease in emissions was recorded for all bundles, with Bundle 4—Scenario 4 emerging as the most effective combination to promote sustainable urban mobility in BMR.

7.2 COSTING OF PMT SERVICES

Public transport in India is considered as a means of transport for the labor class and mostly used by middle and lower class sections of the society and to some extent by school going children due to the subsidy provided to this section of society. This perception needs to be changed to plan and operate services required for different sections of the public living in urban areas. A scientific evaluation of economics of different categories of public bus services helps in proper planning and provision of different categories of public bus services. This could be comprehensively achieved through the various subheads discussed in this chapter. The urban areas in India could be broadly divided into different categories based on the functionality of the city for the purpose of urban bus services.

1. State capitals
2. District headquarters
3. Other business centers in a state

The requirements of bus services are quite different in each of these urban areas. For example, the services in a state capital could be consist of the following:

1. Air-conditioned bus services
2. Express services
3. Limited halt services
4. City services
5. Suburban services

While in the case of the other two categories, the services could be limited to the following:

1. Limited halt services
2. City services
3. Suburban services

Thus, the cost of services needs to be developed in each case separately. What is more important is developing a template for different cost elements in general and evaluating them with current data of urban operations in India. Table 7.1 provides data on different cost elements for selected public bus operations in different cities. It may be noticed that there are wide variations across different cost elements among these cities. For example, personal costs as a percentage of total cost varies from a high of 87.70% in the case of Ahmedabad Metropolitan Transport Service (AMTS) STU to a low of 19.75% in the case of Delhi Transport Corporation (DTC) organization. The same is the case with respect to interest paid to state and central governments as a percentage of total cost from a high of 67.83% of DTC organization to the low of 2.18% of BMTC.

BEST—Bombay Public Transport
DTC—Delhi Transport Corporation

BMTC—Bangalore Metropolitan Transport Corporation
AMTS—Ahmedabad Metropolitan Transport Service

It may be noted from Table 7.1 that revenue in rupees per km operated is highest in the case of DTC, that is, at Rs. 129.69 per km and least in the case of BEST at Rs. 51.72 per km. However, revenue is a function of load factor and the fare structure. The total cost of providing bus service per km works out to a high of Rs. 390.18 per km in the case of AMTS, while it is a minimum at Rs. 65.66 per km in case of BMTC. This wide variation needs a comprehensive review of operations. The cost could be broadly divided into fixed cost and variable cost. Fixed cost includes all the cost items such as salaries, depreciation, office maintenance, and other overheads. Variable costs are the ones which relate to fuel, maintenance, and other operational costs.

The details of cost elements for sample urban public transport services in India at three different points in the past decade is given in Tables 7.2–7.6.

As shown in the tables, the BMTC is the only organization that has been recording profits for many years. However, the same BMTC recorded a loss of 5493.4 crores in the year 2019–20. A main reason for this huge drop in revenue could be attributed to lower ridership due to the COVID-19 pandemic and the growth of metro network in the city. BMTC, in the year 2013, also incurred a loss of Rs. 731.7 million. All the other organizations' trend is a continuous loss, and it increases with time. Thus, there is an urgent need to look into the viability of public bus organizations as a whole and the current tax liability as well as subsidies to be absorbed by these organizations. Any loopholes in the functioning of these organizations should be considered and corrected; otherwise, they will become a great loss to the governments and society at large. On the other hand, if these undertakings do not expand with time and reduce their services based on profitability, public patronage and interest in using them may be affected and the public may shift to personal vehicles.

7.3 MANPOWER REQUIREMENTS IN DIFFERENT CATEGORIES, SUCH AS OPERATIONS AND MANAGEMENT IN BOTH RURAL AND URBAN AREAS

In all major urban areas of India, city bus services are in operation for a period of 14–16 hours a day, then only it is possible to attract more patronage to bus system; otherwise, people may prefer their own personal modes. Operating 14–16 hours a day requires two sets of crews per bus operated. The number of buses operated, and the staff involved in various operations in running the services in urban operations are tabulated below. It may be noted that there is no uniformity in the manpower needs across organizations. As the crew expenditure comprises about 40% of the total expenditure in most of the STUs, the optimum number to be employed is essential. Each organization must work out what is optimum as per their field conditions. It is worthwhile outsourcing any activity which will bring economy to the organization.

Tables 7.7–7.10 provides manpower data of different organizations in India in the last few years. The maximum productivity per person works out to a maximum of

TABLE 7.1

Details of Cost Elements for Sample Urban Public Transport Services in India in 2019

Sl. No.	Description	Urban (India)		BEST		DTC (City)		BMTC STU		AMTS STU	
A: Financial (Rs. in Lakhs)											
A.1	Traffic revenue	633,108		66,963		60,429		180,700		1022	
A.2	Other revenue	365,317						31,296		3029	
3	Total revenue	998,425		83,396		280,780		211,996		4051	
	Revenue Rs. per km	77.9[illegible]		51.72		129.69		52.15		71.63	
B: Costs											
		Value	%	Value	%	Value	%	Value	%	Value	%
	Personal	835,756	41.79	160,618	70.47	172,582	19.75	146,703	54.95	193,56	87.70
	Material	338,576	16.93	39,086	17.15	77,297	8.85	83,561	31.30	1124	5.09
	Taxes	33,478	1.67	5806	2.55	2826	0.32	10,068	3.77	151	0.68
	Interest	623,754	31.19	4963	2.18	592,772	67.83	1416	0.53	751	3.40
	Misc. and others	76,677	3.83	12,972	5.69	11,767	1.34	12,184	4.56	434	1.97
	Payment to HB	44,583	2.23	2040	0.89	525	0.06	--	--	253	1.15
	Depreciation	46,924	2.34	2445	1.07	16,053	1.84	12,996	4.87	--	--
	Total cost	1,999,747		227,931		873,823		266,930		22,069	
	Total cost Rs. per km	156.[illegible]0		141.34		404		65.66		390.18	
C	Surplus before tax	−967,844		−138,729		−590,217		−44,866		−17,866	
D	Profit/loss	−1,001,322		−144,535		−593,043		−54,934		−18,017	
E	% Operating ratio			324		368		128		--	

Source: State Transport Undertakings by Central Institute of Road Transport, Pune 2019–20

TABLE 7.2
Details of Cost Elements for BEST

	2010–11	2015–16	2019–20
Total revenue	111,278	145,378	83,396
Revenue (Rs. per km)	42.55	62.02	51.72
Total cost	149,416	251,570	227,931
Cost (Rs. per km)	57.13	107.33	141.34
Net profit/loss	−38,138	−106,192	−144,535
Profit/loss (Rs. per km)	−14.58	−45.30	−89.63

TABLE 7.3
Details of Cost Elements for DTC

	2010–11	2015–16	2019–20
Total revenue	96,454	100,499	280,780
Revenue (Rs. per km)	33.02	37.59	129.69
Total cost	325,108	570,091	873,823
Cost (Rs. per km)	111.31	213.24	403.60
Net profit/loss	−228,654	−341,110	−593,043
Profit/loss (Rs. per km)	−78.28	−127.59	−273.91

TABLE 7.4
Details of Cost Elements for BMTC

	2010–11	2015–16	2019–20
Total revenue	132,935	220,748	211,996
Revenue (Rs. per km)	29.02	65.68	52.15
Total cost	127,900	219,375	266,930
Cost (Rs. per km)	27.92	65.28	65.66
Net profit/loss	5034	1372.66	−54,934
Profit/loss (Rs. per km)	1.09	0.41	−13.51

TABLE 7.5
Details of Cost Elements for AMTS

	2010–11	2015–16	2019–20
Total revenue	10,890	13,039	4051
Revenue (Rs. per km)	20.74	22.47	71.63
Total cost	24,809	40,690	22,069
Cost (Rs. per km)	47.25	70.12	390.18
Net profit/loss	−13,919	−26,796	−18,017
Profit/loss (Rs. per km)	−26.51	−47.65	−318.55

TABLE 7.6
Comparing the Total Profit and Loss of Different Organization in Past Few Years

Net Profit/Loss (in Million)	BEST	DTC	BMTC	AMTS
2005–06	−2505.776	−6036.008	1023.996	−291.052
2010–11	−3813.825	−20,612.197	499.443	−1211.811
2015–16	−10,619.2	−34,111.0	1372.66	−2679.6
2019–20	−14,453.5	−59,304.3	−5493.4	−1801.7

TABLE 7.7
Urban Undertaking for Staff Positioning BEST

	2005–06	2010–11	2015–16	2019–20
Total staff	34,470	30,183	34,174	27,024
Staff per bus on road	11.21	7.39	8.4	9.01
Manpower productivity (km/staff/day)	19.10	23.74	18.7	16.3
Effective km per crew/day	28.79	31.83		22.4

TABLE 7.8
Urban Undertaking for Staff Positioning DTC

	2005–06	2010–11	2015–16	2019–20
Total staff	28,460	35,557	30,527	28,163
Staff per bus on road	9.07	6.67	6.7	8.74
Manpower productivity (km/staff/day)	25.03	22.52	23.93	22.0
Effective km per crew/day	41.76	30.37		24.8

TABLE 7.9
Urban Undertaking for Staff Positioning BMTC

	2005–06	2010–11	2015–16	2019–20
Total staff	18,150	32,875	35,554	33,292
Staff per bus on road	5.51	5.84	5.5	6.00
Manpower productivity (km/staff/day)	41.59	37.87	25.8	33.4
Effective km per crew/day	51.45	46.60		41.3

TABLE 7.10
Urban Undertaking for Staff Positioning AMTS

	2005–06	2010–11	2015–16	2019–20
Total staff	3886	5274	5498	3546
Staff per bus on road	10.12	7.82	5.5	5.53
Manpower productivity (km/staff/day)	20.39	27.28	28.8	33.9
Effective km per crew/day	31.27	-		41.7

33.9 km/staff/day in the case of the AMTS organization and a minimum in the case of BEST of 16.3 km/staff/day in the year 2019–20.

The above tables give the total staff, staff per bus values, and effective kilometers per employee for different bus undertakings. Every organization presents different staffing rates and in order to give the optimum BSR values, its impact on total cost, passenger satisfaction, and patronage should be given due consideration. A national strategy may be more appropriate in the development of norms for different cost items.

7.4 ESTIMATION OF SOCIAL, ECONOMIC, AND ENVIRONMENTAL BENEFITS ASSOCIATED WITH IMPROVED PMT

There is an urgent necessity in India to provide an efficient public transport system such that the society can derive optimum social and environmental benefits. These two aspects need to be given serious consideration:

1. Optimum fleet size for a given urban area such that more people can use the bus system as shown in Table 7.11.
2. Estimating the economic and social benefits due to that policy.
3. Unfortunately, there is no systematic study undertaken to assess the fleet requirements with relation to population size of an urban area. The limited research undertaken by the author at IIMB provided the guidelines which are given below.

Applying these criteria, the fleet strength needed in different cities of India for three different time periods is presented below. Table 7.12 details the requirement of the number of buses (minimum number and desirable) in the ten most populated cities of India. The requirement of buses for metropolitan cities of India is separately tabulated in Table 7.13. This tables also present the additional fleet needed as of 2020 on the basis of fleet in operation.

A bus on an average provides mobility to about 1000 people in a day and the average distance traveled is about 6–7 km per person. Thus, if the same travel, that is, 6000-to-7000-person-km is to be accomplished by personal mode, the resources (fuel and environmental degradation) spent is three to five times higher than bus travel.

There is no authentic data about the number of buses in operation by private operators in different cities of India. The information presented in Table 7.13 indicates that the current scenario in metropolitan cities is very dismal and needs urgent remedial action. This partly explains the congestion problems of commuters in metropolitan cities. Due to this shortage of bus fleet, many commuters who may be interested in using public transport were forced to use their personal modes.

As per the information provided in the tables above, and the norms of bus fleet based on population shown in Tables 7.11 and 7.12, it is clear that all the bus undertakings are incurring losses. Thus, public bus undertakings have become really white

TABLE 7.11
Norms for Bus Fleet Based on Population

Population Range (Million)	Buses Required per Million Population: Desirable (Scenario 1)	Minimum (Scenario 2)
1–2	400	300
2–4	500	400
4–6	600	500
6–8	800	060
>8	1000	800

TABLE 7.12
Fleet Requirements in Different Urban Centers of India

		2011		2016		2021	
Sl. No.	Name of Cities	Scen.1	Scen.2	Scen.1	Scen.2	Scen.1	Scen.2
01	Mumbai	12,788	10,230	15,984	12,788	23,910	19,128
02	Kolkata	14,456	11,564	180,689	14,456	18,240	14,592
03	Delhi	15,989	12,791	19,987	15,989	21,250	17,000
04	Chennai	7027	5622	8784	7027	11,580	9264
05	Bengaluru	8886	7109	11,107	8886	11,070	8856
06	Hyderabad	8646	6917	10,808	8646	9970	7976
07	Ahmedabad	5649	4237	8827	7061	8270	6616
08	Pune	3521	2934	5868	4401	5264	3948
09	Surat	2636	2196	3295	2746	3582	2985
10	Kanpur	2522	2102	3153	2627	1900	1520

elephants on public money. Both state and central governments stopped funding this organization; thus, the process is really complicated as far as resource mobilization is concerned. However, at the same time, the state government imposes a lot of constraints in their functioning which impinge on their revenue such as:

1. Student concessions
2. Passenger taxes for doing business.

This is best illustrated from the data of APSRTC (now divided into Telangana and Andhra Pradesh) of India which is shown in the below table. The state is divided

TABLE 7.13
Requirement of Buses for Metropolitan Cities in India 2021

Name of City	Estimated 2021 Population (Million)	Buses as of 2020	Bus to Population Ratio	Requirement		Correction for Rail Network	Net Requirement		Deficit	
				Scen.1	Scen.2	%	Scen.1	Scen.2	Scen.1	Scen.2
Ahmedabad (AMTS)	8.27	641	12,902:1	8270	6616	0	8270	6616	7629	5975
Bengaluru (BMTC)	11.07	5542	1997:1	11,070	8856	15	9410	7528	3868	1986
Chennai (CNI)	11.58	3315	3493:1	11,580	9264	30	8106	6485	4791	3170
Delhi (DTC)	21.25	3222	6595:1	21,250	17,000	20	17,000	13,600	13,778	10,378
Mumbai (BEST)	23.91	2998	7975:1	23,910	19,128	50	11,955	9564	8957	6566
Hyderabad (TSRTS)	9.97	2566	3885:1	9970	7976	0	9970	7976	7404	5410
Total	86.05	18,284	4706:1	86,050	68,840		64,711	48,063	46,427	33,485

into three regions. In each zone, the loss-making and the total loss-making routes are presented.

It is impossible for any organization to be in business in such a situation. As a matter of fact, the customer surveys indicate that the people are very happy about the services; thus, in analyzing the functions of bus services, a new criterion called EVA may be adapted. Application for this in the case of APSRTC is presented in the next section.

From Table 7.14 it is clear that the different areas of SBUs show the highest percentage of loss-making routes which is a great loss to the government, and it also failed in the matter of reaching public satisfaction.

7.4.1 Application of EVA Methodology for Evaluating the Financial Performance of the Corporation and Responsibility Centers

In a country like India, the transport undertakings are not allowed to operate buses on cost plus principle, the state government directs them to operate services to the needs of the society irrespective of load factors (demand for services) or return on investment. In addition, the transport services are to provide service at a concessional price to some sections of society especially school going children. As such, evaluating the performance of transport undertaking solely based on financial performance is not proper or desired. A new concept of EVA methodology is suggested in lieu of financial analysis. This concept of EVA is being followed by many corporates. However, this methodology needs to be adjusted to take care of the components of transport operations.

7.4.2 EVA Methodology

The objective of applying EVA is to develop a common financial performance evaluation template which can be used across depots and divisions, as well as organization as a whole. EVA can be calculated as:

TABLE 7.14
Number of Routes and Loss-Making Routes

	2001			2003			2005		
Year	**SBU1**	**SBU2**	**SBU3**	**SBU1**	**SBU2**	**SBU3**	**SBU1**	**SBU2**	**SBU3**
Number of routes	847	383	68	860	310	37	922	270	33
Number of loss-making routes	583	294	64	711	302	97.42	709	270	33
% of loss-making routes	68.83	76.76	94.12	82.67	97.42	100	76.90	100	100
Total % of loss-making routes	79.90			93.36			92.30		

EVA = Net Operating Profit After Tax Capital Charges (Invested Capital × Cost of Capital) EVA formula for APSRTC can be:

Profit + Opportunity Cost of Concessions + Differential MV Tax Adjustment

Less (Opportunity Cost of Equity by Government and Actual Interest Charges)

EVA is used in the private sector to make adjustments to profit and loss to adjust for accounting. In a passenger road transport system context, it can be used for incorporating the opportunity cost of meeting socioeconomic services. Here the profits can be adjusted for an unreimbursed amount of concessions, differential MV taxes, and the cost of providing uneconomical services.

1. The cost of capital of both equity and borrowings is then deducted from the adjusted profit to arrive at EVA.
2. Deducting the cost of capital will make the responsibility center managers conscious of asset cost.

EVA enables financial comparison at division and depot levels as the impact of external factors are ironed out. The purpose of this application will be to evaluate if the operations are efficient after meeting these obligations. The responsibility center in-charge cannot take refuge under the argument of incomparability. It also helps to point out the corporation's contribution to other sectors in financial terms.

7.4.3 EVA Computation

EVA valuation of APSRTC for the years 2000–01 to 2005–06 are provided (Tables 7.15–7.17).

Three adjustments were considered:

1. Unreimbursed amount of concessions
2. Difference between actual MV tax paid and benchmark rate of MV tax of 6.25%
3. Losses from providing interior rural routes

Valuation by two methods is provided:

1. After adjusting for only the unreimbursed amount of concessions and MV tax differential
2. After adjusting for the above plus losses incurred on interior rural routes.

7.4.4 EVA Results—Observations

1. It can be seen from the tables that EVA helps to capture the net impact of external factors.
2. It can be observed from the table that even though profits have been fluctuating, EVA has been showing a steady trend and is significantly positive.

TABLE 7.15
EVA Method 1

Years	Profit	Unreimbursed Student Concession	Differential MV Taxes	Adjusted Earnings	Capital Charges (Million)	EVA
2000–01	–2099	2028	2157	2086	834	1252
2001–02	–2720	1512	1867	659	1071	-412
2002–03	–1817	1398	1567	1148	1175	-27
2003–04	–420	1520	1660	2760	1253	1507
2004–05	–2248	1675	1783	1210	1231	-21
2005–06	–521	0	718	197	1120	-923

TABLE 7.16
EVA Method 2

Year	Profit	Interior Rural Losses	Unreimbursed Student Concession	Differential MV Taxes	Adjusted Earnings	Capital Charges (Million)	EVA
2000–01	–2099	2028	1508	2157	4225.46	834	3391
2001–02	–2720	1512	1121	1867	2648.86	1071	1578
2002–03	–1817	1398	1724	1567	3845.74	1175	2671
2003–04	–420	1520	1694	1660	5504.95	1253	4252
2004–05	–2248	1675	2072	1783	4310.72	1231	3080
2005–06	–521	0	1948	718	3063.37	1120	1943

TABLE 7.17
EVA: Hyderabad City Region

Year	Profit Before Interest	Unreimbursed Concession	MV Tax Adj	Adj Profit	Total Assets	Capital Charge (int @ 8%) in 0.1 Million	EVA
2000–01	–1487	2839	2895	4247	30,608	2448	1799
2001–02	–1394	2737	2424	3767	30,282	2422	1345
2002–03	901	3716	1671	6288	28,842	2307	3981
2003–04	1261	4489	1827	8677	31,306	2504	6173
2004–05	2332	4626	1961	8919	33,245	2660	6259

3. In the last year, EVA has fallen even though operational performance improved because of oil price hike and salary hike.

7.4.5 EVA Results—Hyderabad City Region

1. The templates can be extended to SBU levels and depot levels.
2. A model computation undertaken in Hyderabad city region is provided for illustration.
3. For Hyderabad, profits were adjusted for unreimbursed amount of student concessions and MV tax differential.
4. Buses with Re.1 written down value were valued at saleable value and land at guideline value.
5. Interest charges are assumed at 8%.

7.4.6 Financial and Economic Scenario

The revenue and profitability of any bus undertaking is influenced by many factors. What is really needed is a template combining all these factors over several time periods and changing each of the variables will estimate the financial position of the organization. A macro developed by IIM for APSRTC is used for purposes of illustrations and from that it is clear, if the organization can increase the load factor by a small value over a number of years and reduces the bus staff ratio to some extent, it would result in profitability as well as wiping out the accumulated losses.

This macro will provide areas of improvement and its impact on profitability. Thus, if the authorities bestow attention, it is not difficult to make their profit over a five-year period. It is ascertained by the concerned officials that the load factor and staff ratio values are achievable. Some undertakings were able to reach these targets. Thus, what is lacking is the will to achieve these targets.

7.4.7 Environmental Benefits of Public Transport

7.4.7.1 Urban Passenger Transport

In the case of urban passenger transport needs, the current scenario of Bengaluru is discussed in detail. Detailed information for other cities is not well documented for undertaking a comprehensive analysis. The trip rates could be used for assessing the needs but not to identify the locations where specific improvements are needed to avoid congestion and minimize environmental problems.

7.4.7.2 Current Scenario in Bengaluru

Bengaluru has become a true cosmopolitan city and is witnessing tremendous activity in industry, trade, and commerce, leading to an exponential growth in population. Bengaluru had a population of 5.7 million people as per 2001 census records, 8.4 million as per 2011 census records, and an estimated population of more than 13 million in 2023. This unprecedented growth has created severe strain on the infrastructure like the urban water and sewerage system, transport network, telecommunication systems, etc. The public transport needs are met by BMTC largely but not to the satisfaction of the growing needs of commuters.

The road network of any city consists of arterial and nonarterial roads. Arterial roads are normally designed to the standards of state highways. Bengaluru has a total network of 14,000 km of road out of which only around 920 km are arterial roads. National and state highways entering the city add to the arterial roads carrying most of the vehicular traffic. The road networking has not been scientifically planned, as the city did not expect the unprecedented growth of population and economic activity. There is no provision for expansion or for widening of the existing roads. The recent development of outer, inner, and satellite ring roads are expected to provide some respite to the current traffic woes. Traffic planning is left with only a choice of one-way systems. The construction of underpasses and flyovers may provide some relief to the high-density traffic of city roads. The present road network consists of ring roads and major radial corridors. A number of proposals have already been included in the Comprehensive Mobility Plan (CMP) for the city. It is necessary to integrate/superimpose all these proposals in the light of projected travel demand for road traffic and in conformity with each other and there is neither conflict nor duplication. For decongestion of traffic in Bengaluru city, the government has been constructing flyovers and new roads and expanding the existing infrastructure (wherever possible). However, these measures seem to have little impact on actually decongesting the city. Baseline environmental studies are conducted to assess the current parameters.

1. Ambient air quality
2. Water quality
3. Identification of project affected persons (PAPs)
4. Soil erosion details
5. Noise levels, etc.

The urban air pollution is contributed generally by a variety of sources such as industrial, commercial, and transportation sectors. However, in Bengaluru, air pollution problems which are quite severe are mainly compounded by the transportation sector while the other sources such as industrial are contributing less.

A study conducted by the Karnataka State Pollution Control Board (KSPCB) in 2019 revealed that emissions from the transportation sector, including tailpipe emissions (51%) and re-suspended dust (17%), contributed to 68% of total PM10 emissions and 60% of NOx emissions. Notably, just 3% of commercial vehicles were responsible for 84% of PM10 emissions from transportation, while diesel cars accounted for 9%. In terms of sectoral contributions, the transportation sector was found to be a major source of pollution, accounting for 55.3% of the total annual PM10 concentrations and 67.2% of PM2.5 concentrations in 2019. Road dust, including re-suspended particles, contributed an additional 23.8% to PM10 and 8.3% to PM2.5 concentrations annually. The study projected that under a BAU scenario, the total PM10 emission load for the BBMP would rise to approximately 28,000 tonnes per year by 2024—a 15% increase compared to 2019 levels. Transportation and road dust are expected to remain the primary sources of pollution. Promoting affordable and clean fuel-based public transport, such as CNG or EVs, with seamless end-to-end connectivity could significantly reduce reliance on private vehicles and help lower emissions from the transportation sector. As the core transportation sector presently consists mainly of petrol- and diesel-driven vehicles operating throughout the city, the major air pollutant components are contributed by the automobile exhaust emissions, which consist of suspended particulate matter (SPM), respirable suspended particulate matter (RSPM), oxides of nitrogen and sulfur, carbon monoxide, etc. and are monitored by the KSPCB at some locations and by the Central Pollution Control Board (CPCB) at some selected important intersections of the city. The KSPCB during the year 2017–18 monitored 16 locations including two Continuous Ambient Air Quality Monitoring Stations (CAAQMS). The board recommends that the air

TABLE 7.18
Air Quality Levels at Selected Metro Corridor Locations

Sl. No.	Location	Air Quality Index	Result
1	Bangalore University	25	Good
2	CAAQMS at S. G. Halli	50	Good
3	Urban Ecopark Peenya	95	Satisfactory
4	Swan Silk Pvt. Ltd. Peenya	97	Satisfactory
5	Yeshwanthpura Railway Station	95	Satisfactory
6	AMCO Batteries Mysore Road	86	Satisfactory
7	Banaswadi police Station	69	Satisfactory
8	Madavachari House, [illegible]	69	Satisfactory
9	UVCE K. R. Circle	71	Satisfactory
10	Victoria Hospital	67	Satisfactory
11	Indira Gandhi Child Health Care Centre	66	Satisfactory
12	Export Promotional Park, ITPL Whitefield	103	Moderate
13	Rail Wheel Factory, Yelankha	119	Moderate
14	Central Silk Board, Hosur Road	113	Moderate
15	CAAQM City Railway Station	101	Moderate
16	TERI Office Domlur	112	Moderate

quality index less than 50 is permissible; 51–100 is satisfactory; 101–200 is moderate; 201–300 is poor; 301–400 is very poor; and above 400 is considered severe air pollution. The air quality index at the 16 monitored locations is presented in Table 7.18.

7.4.7.3 Recommendations and Action Plan of the Expert Committee on Auto Fuel Policy, Government of India

1. Reducing idling, stop time, and speed changes will considerably reduce pollution and improve the environment.
2. Develop a strategy to discourage the use of private vehicles (cars and two-wheelers) by levying a fee as well as for parking. The charges should reflect the true costs involved in traffic and transportation activities. This may positively influence the demand for public transportation.
3. Road pricing must reflect the true costs. This may reduce peak hour. The charges for using the roads during peak hours could be higher by two to three times than those other hours of the day.
4. Physical restrictions could be imposed on entry of vehicles either during part or on the whole of the day into certain areas.

SUMMARY

This chapter discusses the economic condition of the STUs of different states in terms of total revenue, different types of transport taxes, operating costs, and the net profit/loss. It also deals with the manpower requirements like total staff ratio with regard to the number of buses, total road length, and the population and how it effects the economic stability of STUs. This chapter also gives information about the bus fleet for populations of the different cities and discusses the EVA methodology for financial evaluation. Mainly it shows some highlights of environmental issues like emission standards and air quality index.

8 Resource Needs and Organizational Structure for Public Mass Transport at State Level

8.1 CURRENT FINANCIAL PERFORMANCE OF PMT ORGANIZATIONS IN INDIA

The financial performance of any public transport undertakings (metro, public bus systems) depends upon the revenue it generates on the basis of operations and how much it needs to be spent on various inputs in running the organization. The revenue stream of bus undertakings mainly includes two sources, namely ticket sales and advertisement revenue on buses and other bus-related infrastructure. However, the major share is contributed by ticket sales. The cost stream includes capital costs associated with buses or metro rail coaches, maintenance, fuel costs, and manpower costs. In addition, the organizations incur expenditure in creating infrastructure like bus stands, stops, metro stations in the case of metro system, etc. In India, for quite some time, the Government of India and respective state governments provided some financial assistance to consolidate the working of these organizations. In India, it is normal practice that whenever some financial assistance is provided a number of social obligations are also tied with it. Bus undertakings are no exception to this phenomenon. Running buses in cities with very low demand during off-peak hour and student concessions are some of the examples of this phenomenon. For sustainable growth in the fleet to take care of demand increase and replacement needs of old stock, the undertakings must enjoy surplus revenue or profits. The following paragraph details the financial performance of bus undertakings in India.

8.1.1 Financial Performance: Turnover of STUs

Even though the PMT is provided by both the metro and public buses, the information is mostly available in the public domain with respect to bus system. As metro operations and other related information is not easily available in detail, we restrict ourselves to bus operations only.

The financial performance of STUs does not truly reflect their physical performance. Currently, the total revenue earned by them is in the range of about Rs. 566,662.7 million per year on which they make a loss of Rs. 197,262.6 million per annum. Some STUs like Karnataka SRTC, Bangalore Metropolitan Transport

DOI: 10.1201/ 9781003740247-8

Corporation (BMTC), UPSRTC, some Tamil Nadu STUs, etc. have been showing profits in recent times. Also, STUs have been consistently adding to physical infrastructure in spite of poor earnings. Even though the capital base of STUs is estimated to be around Rs. 592,235 million, the value of the assets owned by all the STUs in the country may exceed about Rs. 592,853.5 million taking into account the rolling stock, the lands, buildings, and other items of infrastructure owned by them.

The poor financial performance of STUs despite their robust physical performance is more due to the impact of rising input prices relating to MV taxes, concessions and universal service obligations, unviable fare structures, and fuel prices. The following paragraphs analyze these factors. The main causes are as follows:

1. MV taxes
2. Cost of concessions
3. Uneconomic fare structure

8.1.1.1 MVs Tax

STUs pay substantial amounts of MV taxes to state governments, usually as a percentage of traffic revenue. The nomenclature, incidence, and method of calculation of the taxes vary from state to state. The amount of MV taxes paid over the years by STUs as a group at the all-India level is shown in Table 8.1. The total tax paid in 2020 by STUs for which data is available was about Rs. 12,655.8 million which represented 2.23% of their total revenue.

Thus, it may be seen that the states have usually treated STUs as an important source of revenue while at the same time expecting them to provide inexpensive transport services to the people. This runs counter to public policy prescriptions of encouraging the development of affordable public transport. The STUs are under the administrative control of respective state governments.

A bus during its lifetime (over a ten-year period) pays a much higher amount than any other mode of transport towards tax. It is a matter of concern that environmentally friendly and congestion-reduction oriented transport should pay higher taxes than others.

TABLE 8.1
MV Tax—All India

Year	Total Revenue (in Million Rs.)	MV Tax (in Million Rs.)	MV Tax Rate %	Losses of STUs (in Million Rs.)
2016–17	559,024.9	13,047.5	2.33%	171,686.0
2017–18	593,652.1	13,791.1	2.32%	203,095.7
2018–19	645,543.6	14,218.1	2.22%	179,239.2
2019–20	566,662.7	12,655.8	2.23	197,262.6

Source: State Transport Undertakings from 2016 to 2020

8.1.1.2 Cost of Concessions

Another factor that significantly affects the revenues of STUs is concessions granted by state governments to various categories of users of STU services. According to statistics available with the CIRT, the cost of these concessions to the STUs amounted to Rs. 1.137×10^4 million in 2000–01 and Rs. 1.482×10^4 million in 2004–05. These concessions have increased to 7.34×10^4 million in 2020. While some states reimburse the losses from concessions to STUs partly, in most cases such compensation by way of reimbursement is grossly inadequate. Financial burden, due to concessional fare and free passes as announced by the government, becomes inevitable to STUs. Though STUs deserve complete reimbursement of such burden, only few state governments grant subsidies to compensate STUs in this regard. It should be mentioned here that while the STUs are expected to provide affordable transport to students they cannot be expected to bear the full burden of the costs.

8.1.1.3 Uneconomic Fare Structure

The rising operational costs, taxes, and the burden of concessional travel schemes introduced by state governments are seldom adequately compensated in the fare structure of STUs which is determined by the state governments. For political and other considerations, the state governments are reluctant to allow STUs to revise the fares with the periodicity that it requires. This situation was noted by the National Transport Policy Committee as far back as 1980 when they noted that

> many State Transport Undertakings are presently operating at a loss, mainly on account of uneconomic fares which have been kept low as a deliberate policy of Governments. Such low fares are not in conformity with the principle of covering short term marginal costs. We therefore strongly recommend that fare structure of State Transport Undertakings should be revised and brought in line with costs structure. We also urge that within the broad policy frame which may be laid down by the proposed National Transport Commission, each Public Sector Undertaking should have the freedom and flexibility to revise and adjust its fare structure.

Thus, the fare structure of STUs in most cases is always found to be lagging their operational costs, thereby forcing the STUs to operate at negative margins. This is although the operational costs of STUs were increasing at a much higher rate than the consumer price index or wholesale price index mainly on account of the steep increase in the prices of HSD. Therefore, the EPKM of STUs as a group were lower than their cost per 317ilometre (CPKM) as shown in Table 8.2. The earnings mentioned in Table 8.2 include the subsidies and other non-traffic revenues that STUs receive. The situation is even worse if the sources of non-traffic revenue are removed.

In addition, STUs expected to operate on interior rural routes, hilly, and remote areas which are highly subsidized. The STUs seek to cross-subsidize such operations with the revenue generated on bus routes with better load factor. The scope for cross-subsidization got eroded as the state governments permitted private operators to operate in the profitable routes with no compensating service obligations.

TABLE 8.2
Profit/Loss per Kilometer (All India)

Years	EPKM (ps per km)	CPKM (ps per km)	Profit/Loss (ps per km)
2006–07	1836.1	1923.9	−87.80
2016–17	3347.3	4375.3	−1028.0
2017–18	3544.6	4757.3	−1212.7
2018–19	3894.1	4975.3	−1081.2
2019–20	3955.0	5332.0	−1377.0

Source: State Transport Undertakings from 2006 to 2020

TABLE 8.3
Personnel Costs in STUs

Type of Operations	Personal Costs (in Million Rs.)	Total Costs (in Million Rs.)	Share of Personnel Costs (%)
2016–17	322,170.1	730,711.0	44.08%
2017–18	358,804.5	796,748.1	45.03%
2018–19	350,413.8	824,782.9	42.48%
2019–20	342,442.1	763,725.2	44.84%

Source: CIRT, Pune

8.1.2 Manpower Costs in STUs

STUs, being seen as state's own agencies, also employ their workers on terms similar to those of government employees besides complying with the provisions of law relating to welfare of labor. The personnel costs of STU operations have steadily increased over the years from 395 ps per km during 1995–96 to 687 ps per km in 2004–05 to 2646 ps per km in 2020. This increase is mainly the result of the periodical revision of wages to the workers, which is necessitated when the state governments revise the rates of dearness allowance to their employees. The personnel cost across different sectors of rural, hilly, and urban transport services is presented in Table 8.3.

Despite the frequent wage revisions, the proportion of personnel costs in the total costs of STUs has almost remained same over the years. The share of personnel costs in the total costs at the all-India level was 42.4% in 2000–01, 39.7% in 2004–05, and 44.84% in 2019–20. In spite of the periodical wage increments for staff, the personnel costs as a proportion of total costs have almost remained the same due to productivity improvement of STU fleets as mentioned above and due to the reduction achieved in the staff per bus ratio. It is also partly due to a steep increase in fuel costs in recent years, which has increased the share of fuel costs relative to personnel costs.

8.1.3 Replacement and Augmentation of Rolling Stock

The cost of rolling stock, spare parts, tires and tubes, etc. has been steadily increasing over the years. Given the operating conditions in India, the buses need to be replaced over a much shorter period than is the case in other countries. The STUs therefore find it necessary to replace about 12 of their rolling stock every year to ensure that the overaged vehicles are phased out from their fleet. Further, there is also the need to augment their rolling stock at the rate of at least about 10% every year to meet the demand for increased passenger services. Thus, the STUs need to acquire new rolling stock every year to the extent of about 20–25% of their current fleet strength if they are to provide satisfactory service. The demands of long-distance bus users and the urban commuters for buses of superior design and comfort have made it necessary for the STUs to invest substantial amounts to acquire new rolling stock of better quality.

The above factors show that the STUs need to invest in acquiring about 25,000–30,000 new buses every year to sustain efficient operations. These resources have to be generated annually out of their depreciation reserves and profits. A significant part of the cost of the rolling stock is accounted for by central and state taxes. The excise duty levied by the central government on the bus chassis and the sales tax and GST levied by the state governments on the chassis as well as the bus bodies amount to a major chunk of the total cost of a bus acquired by an STU.

8.1.4 Investment in Infrastructure

While the STUs operate only about 20–25% of the bus services in most states (in states like Andhra Pradesh, Tamil Nadu, Karnataka, and Maharashtra, STUs operate more than 50% of bus services), they are involved in providing the bulk of infrastructure facilities like bus stations, bus shelters, and other passenger amenities. The STUs are usually required to meet these obligations out of the revenues generated by their operations. In addition, the STUs have also to invest on the construction of bus depots in different parts of the state and provide workshop facilities in order to be able to provide reliable transport services to the people. This is in contrast with the private bus operators who are not expected to invest on bus stations, bus depots, etc. owing to the small size of their fleets in most cases. While there are no reliable estimates of the investments made by STUs in respect of infrastructure, it may be reasonable to assume that such investments account for at least 20% of the value of the rolling stock owned by the STUs. The STUs in many cases have borrowed funds or used their own revenues for making such investments on infrastructure, and they incur substantial debts on such investments made for the benefit of the traveling public.

8.2 REVIEW OF THE EXISTING FUNDING PATTERN

The STUs mostly draw funds from the respective governments and to an extent from commercial banks.

8.2.1 Capital Structure and Investments of STUs

These organizations have been investing in infrastructure despite the absence of a clear policy to strengthen the capital structure of the STUs. They have not been augmenting their buses, but they have been trying to upgrade their bus stands due to political compulsions. The more profitable STUs are in the forefront of modernizing their fleets, which is indicative of a strategy for improving profitability. It is also noteworthy that the STUs have been consistently providing for depreciation reserves of substantial amounts.

8.2.2 Capital Contribution to STUs

In the wake of the formation of state road transport corporations, the central government introduced a provision for making capital contributions to the STUs to strengthen their capital base. Under the scheme, the state governments also had to extend corresponding budgetary support toward the equity of the corporations. There was no capital infusion coming after the initial burst of contributions. Unfortunately, the state governments too found it convenient to discontinue their capital contributions to STUs. This, coupled with the losses sustained by the STUs due to the factors discussed above, completely eroded the capital base of STUs in the country. The STUs carry huge accumulated losses on their balance sheets, and they are managing the deficits with borrowed funds and compensation from state governments. Borrowing funds by the STUs with their negative net worth has proved to be difficult and expensive. The combined interest of all the STUs on outstanding borrowings for the year 2019–20 stood at about Rs. 82,451.4 million.

8.3 ESTIMATION OF RESOURCE REQUIREMENTS AT STATE AND CENTRAL LEVEL FOR THE NEXT TWO DECADES

The overall financial implications for the center, states, and STUs are discussed in this section. This follows from the discussion in the previous chapters on demand forecast and policy initiatives prescribed to enable the states to meet demand.

8.3.1 Infusion of Capital

The first step toward revitalizing STUs will be to create clean balance sheets. STUs carry huge accumulated losses on their balance sheets for reasons discussed so far. These losses are more than their net worth, and this seriously erodes their financial strength. It makes borrowing much more difficult. Government as promoters needs to inject funds to inspire confidence among banks and financial institutions. Infusion of equity capital in the form of capital contribution from the state and the central governments is a measure, which needs to be initiated immediately. The earlier plan of capital contributions to STUs should be resumed immediately. The capital infusion needs of each STU should be assessed keeping in view their long-term viability. This would enable the state governments to prepare a long-term investment plan for each of the STUs and provide the required funds accordingly. This can be achieved by handling the debt burden of the STUs.

8.3.2 Policy Toward Debt Burden

The STUs have been primarily financing their deficits through borrowings in the absence of support from governments. This has now grown to be quite huge and has reached the stage when most of their borrowings go into debt servicing itself. This cycle has to be broken. Debt servicing is one of the important items of cost for STUs after considering major costs like fuel and personnel. It is therefore necessary to reduce the debt burden of the STUs.

It is recommended that 50% of the outstanding loans of the STUs should be taken over by the central and the state governments. Financial relief through debt taken over by the government has in the past been given as in the case of state electricity boards to enable them to participate in a program of power sector reforms. The balance of 50% would be serviced by the STUs.

8.3.3 MV Tax

During 2019–20, the reporting STUs paid Rs. 41,219.1 million as taxes to the exchequer, which is 7.3% of their total revenue. Rural STUs paid 8.2% of total revenue as tax, while hill region STUs paid 5.4% and urban STUs paid 3.4% of their total revenue as tax. States see this as a major source of revenue. Therefore, the center should consider compensating the states for the loss for the first three years, and the states should bear it in the subsequent years. The collections will go up substantially given the buoyancy of this sector.

8.3.4 Students' Concessions

The implications of students' concessions have been discussed in the previous sections. It has also been suggested under the policy section that the fare should be equally shared by the users (students), STUs, and state governments, as this is primarily a state government's decision.

8.3.5 Universal Service Obligations

Service obligations in terms of uneconomic services like rural and urban services need to be spelled out. It is expected that 20–30% of routes are expected to fall under this category. STUs or private operators should be given benchmarks for cost per kilometer and EPKM, and the balance will be met through compensation from a pool. This pool will be created through cess or deficit charges on the lines of telecommunications, from luxury services, and other profit making routes or sectors. If the states are unwilling to do this, it should be prepared to meet the deficit. States should take steps to reduce their share of the number of these routes.

8.3.6 Fleet Augmentation

In the section on demand forecast, the total number of additional buses is estimated to be around 130,000. This would require an outlay of Rs. 20,000 million over the five-year plan. It is recommended the center should bear 50% of this burden apart from reducing duties on bus chassis. States should bear 25% of the investment and also reduce sales tax on spare parts and bus body building. STUs should bear the balance of the 25% investment. Enable the STUs to meet about 50% of the cost of the new buses to be procured to replace their overaged rolling stock. The STUs will raise the balance of 50% by borrowing from institutional sources or by using their depreciation reserves.

8.4 IDENTIFICATION OF INNOVATIVE STRATEGIES TO GENERATE REVENUE IN A SUSTAINABLE MANNER

There is an urgent need to market the concept of effective and efficient public bus systems in all million-plus cities to minimize environmental degradation, conserve the usage of petroleum products, and reduce traffic congestion, especially in metropolitan cities. It is well known that if a bus runs half empty (load factor 50%) the revenue for the empty seats is lost once for all. Load factors could be improved through scientific study of demand and reviewing the current routing and scheduling once in two years to enhance ridership. It is common knowledge that during off-peak hours the load factor will be definitely less and adjust the fares accordingly to meet at least the operating cost during off-peak hour. Another innovative way is to provide a variety of services (air conditioned, limited halt, express, and deluxe services) during peak hours with different fare structures to cater to the needs of different classes of people. In India, there is a perception that bus transport is meant for workers, schoolchildren, housewives, and people belonging to low-income groups. This perception needs to be changed through the provision of different types of services mentioned above. In order to retain the ridership some sort points (similar to frequent flier mileage points of airline companies) may be followed. In order to encourage the use of public transport all the organizations employing more than 250 people may be persuaded to contribute annually at the rate of Rs. 250 per employee. The governments need to contribute to equity capital of these undertakings at least to the tune of concessions provided by these organizations. For all the monthly pass holders in addition to concessional fare accident insurance, reimbursement of medical expenses due to bus accidents could be conceived and marked vigorously. All these strategies improve the ridership and revenue for the organization. It has been noticed that decrease in congestion normally brings in the reduction in accident rates and consequently the outgo from insurance companies. They may be persuaded to part with 1–2% of revenue annually for improving the public transport system in urban agglomerations.

8.5 THE NEED FOR DEDICATED FUNDING AS A PERCENTAGE OF STATE REVENUE AND CREATION OF STATE-LEVEL ORGANIZATION FOR OPTIMUM USAGE OF THESE RESOURCES

Public transport plays a crucial role in ensuring affordable and accessible mobility, reducing emissions, and supporting economic growth. However, the sector continues to face financial challenges due to inconsistent funding and fragmented governance. To meet social objectives such as emission reduction, equitable transport access, and sustainable urban development, a comprehensive restructuring of funding mechanisms and the establishment of a state-level organization are essential.

Transport has two major components: economic development and service provision. All modes of transport fulfil these two components but vary widely in their balance. For example, air transport is predominantly an economic activity (90% economic component and 10% service component), while railways balance both at approximately 50% each. In contrast, public bus transport primarily serves the public good, with 90% of its role being a service component and only 10% contributing to economic development. When the economic component is higher, users can be charged based on the cost-plus principle. However, in cases where the service component is dominant, such as public bus transport, fares cannot be fully charged to users, and significant subsidies are required to ensure accessibility for all.

Unlike other essential public services such as water supply, education, and healthcare, transport funding is often subjected to strict financial and economic evaluations. However, public transport should be recognized as a public necessity rather than a commercial entity expected to generate profits. Sustainable transport systems require long-term financial support, ensuring affordability and efficiency for all users.

To achieve this, a fixed percentage of state revenue should be allocated to public transport, providing a stable financial base. Current reliance on irregular budgetary allocations leads to service disruptions, delayed infrastructure projects, and poor maintenance. Establishing dedicated funding mechanisms—such as fuel taxes, congestion pricing, and land value capture—can ensure continuous financial support for public transport improvements.

Additionally, initial capital investments for transportation systems, including buses, metro systems, and railway expansions, should be fully funded by state and central governments. Many large-scale transport projects, such as metro systems and railways, already rely on government support, yet they struggle to recover operational costs. Without adequate funding, even well-planned transit systems fail to operate efficiently, affecting public usage and long-term sustainability.

For the first five to ten years, operational losses of public transport systems should be absorbed by the government and society. Public transit systems take time to develop a steady ridership and become financially sustainable. Subsidizing operational costs in the initial years ensures affordable fares, improves service quality, and encourages more people to shift from private vehicles to public transport, reducing congestion and emissions.

Furthermore, many state governments have recently introduced free travel for women in addition to existing concessions for various sections of society. While such initiatives enhance accessibility, they make it even more challenging to generate funds for expanding bus fleets to meet rising demand and replacing aging vehicles. Given these social obligations, it is unrealistic to expect public bus services to be self-sustaining. State and central governments must recognize these financial realities and allocate sufficient funds to ensure the continued provision of essential transport services.

Alongside financial reforms, a state-level transport organization is necessary to streamline planning, funding allocation, and resource utilization. Currently, multiple agencies oversee different aspects of transportation, leading to inefficiencies and a lack of coordination. A centralized authority at the state level can ensure better integration of transit services, improve efficiency, and optimize financial resources.

SUMMARY

This chapter mainly highlights the financial issues of public transport organizations, in terms of MV tax, fuel cost, fuel efficiency, rolling stocks, turnover of STUs, and mainly investment in infrastructure. It also throws some light on the existing funding pattern of STUs and how these terms affect the running status of the government organization.

References and Bibliography

Anantharamaiah, K.M. and Ramanayya, T.V. (1991). Rural travel characteristics a case study of Ratnagiri district. *Moving Technology, Journal.*

Anantharamaiah, K.M. and Ramanayya, T.V. (1992). *Changes in Rural Travel Characteristics in Southern India Published in the World Conference on Transport Research.* Lyons, France.

Anantharamaiah, K.M. and Ramanayya, T.V. (1994a). Characteristics of vehicle ownership in rural areas, *Indian Highways. Journal of Indian Roads Congress*, New Delhi.

Anantharamaiah, K.M. and Ramanayya, T.V. (1994b). Rural passenger travel characteristics—A case study of southern India. *Traffic Engineering, Journal of Indian Roads Congress*, New Delhi.

Anantharamaiah, K.M. and Ramanayya, T.V. (1994c). Settlement connectivity and rural travel characteristics a case study of southern India. *Journal of Institution of Town Planners*, India.

Assessment of Social and Economic Impact of Rural Roads Component, Sponsored by Government of Andhra Pradesh, May 2007, unpublished, IIM Bangalore.

Bangalore Metro Rail Corporation Limited. Available at: https://english.bmrc.co.in

BDA. (2020). *Master Plan for Bangalore 2031.* Bangalore Development Authority, Government of Karnataka.

Buchan, K. (1992). Enhancing the Quality of Life. In Roberts, J., Cleary, J., Hamilton, K. and Hanna, J. (eds.), *Travel Sickness: The Need for a Sustainable Transport Policy for Britain.* Lawrence and Wishart.

Buehler, R. and Pucher, J. (2011) Making public transport financially sustainable. *Transport Policy*, 18(1), 126–138.

Casey, Jr, H. J. (1955). The law of retail gravitation applied to traffic engineering. *Traffic Quarterly*, 9(3), pp. 313–321.

Census of India. (2011a). Available at: http://www.censusindia.gov.in/

Census of India. (2011b). RITES (2011); LDA (2005); MP Town & Country Planning Office (2008); Guwahati City Development Plan (2006); Government of NCT of Delhi (2009); Government of Uttar Pradesh (2006); Business Standard (2013); Directorate of Economics & Statistics, Karnataka; Directorate of Economics & Statistics, Uttar Pradesh; Directorate of Economics & Statistics, Madhya Pradesh; Directorate of Economics & Statistics, Assam (2013).

Central Institute of Road Transport. (2000–2011). *State Transport Undertakings—Profile and Performances*, Central Institute for Road Transport, Pune, India

Cervero, R. (1998). *The Transit Metropolis, A Global Inquiry.* Island Press, Washington.

Chakroborty, P. Recent Trends in Sustainable Transportation Planning, Lecture Notes, Jointly organized by CiSTUP and Dept. of Civil Eng., IISc Bangalore, 12th to 14th December 2009.

Chennai Unified Metropolitan Transport Authority. Available at: https://cumta.tn.gov.in

CIRT. (2007). Physical Performance of SRTU's for the year 1995–2007. *Indian Journal of Transport Management*, Central Institute for Road Transport, Pune, India

CIRT. (2011). Delhi Metro Rail Corporation Limited; Bangalore Metropolitan Transport Corporation; Uttar Pradesh State Road Transport Corporation; RITES (2012); Assam State Road Transport Corporation; RITES (2011b).

CIRT. (2023). Review of the performance of SRTU's for the year 2016–2023. *Indian Journal of Transport Management*, Central Institute for Road Transport, Pune, India

Consulting Engineering Services India, Ltd. (CES). (2001). Proposed mass rapid transit for Thane city, Draft Final Report, MSRDC, Mumbai, India.

Daly, A. (1975). *Measuring Accessibility in a Rural Context. Rural Transport Seminar*. Polytechnic of Central London.

Delhi Metro Rail Corporation Limited; Bangalore Metro Rail Corporation Limited; Central Road Research Institute, Delhi; ifmo (2013).

Directorate of Economics & Statistics, Delhi (2009); RITES (2011); Lucknow Municipal Corporation, DUDA; DMG Consultancy Services (2011); Guwahati Metropolitan Development Authority.

District level study of socio economic aspects of rural roads in the development of standards for planning rural roads in India (Ratnagiri in Maharashtra State) project sponsored jointly by Planning Commission, Ministry of Shipping and Transport, Government of India, and Maharashtra Government 1981–83, unpublished, IIM Bangalore.

DMRC. (2003). *Detailed Project Report—Bangalore Metro (Phase I)*. Delhi Metro Rail Corporation Ltd., New Delhi, India.

Enoch, M., Potter, S. and Ison, S. (2005). Strategic approach to financing public transport through property values. *Public Money & Management*, 25(3), 147–154.

Furness, K.P. (1965). Time function iteration. *Traffic Engineering and Control*, 7(7), 458–460

GOI. (1987). *Report of the Study Group on Alternative Systems of Urban Transport*. Government of India, New Delhi, India.

GOM. (2000). *Mumbai Sky Bus Metro: Preliminary Engineering Cum Survey Report*. Konkan Railway Corporation, Ltd., Government of Maharashatra, India.

Gould, P. (1969). *Spatial Diffucion*. Commission on College Geography, Association of American Geographers, Washington DC.

Gray, G.E. and Hoel, L.A. (1979). *Public Transportation: Planning, Operations, and Management*. Prentice-Hall Inc., Englewood Cliffs, New Jersey.

Harman, L. (1998). *Technological Change and Public Transportation—Issues for the Year 2020*. Draft Manuscript, Massachusetts Executive Office of Transportation and Construction, Boston, Massachusetts.

Hillman, M. et al. (1973). *Personal Mobility and Transport Policy, Political and Economic Planning*. Broadsheet 542, London.

Hills, P.J. (1996). What is induced traffic? *Transportation*, 23(1), 516.

http://www.censusindia.gov.in Census of India. (2001). Government of India Ministry of Statistics and Programme Implementation Central Statistics Office Social Statistics Division. Census 2001.

http://www.Indiandemographics.weebly.com/settlement-patterns

http://www.pmgsy.in (PMGSY site (project details state wise report) (Last accessed 30.10.2013).

http://www.mapsofindia.com

http://www.gdpofindia.com

http://www.azadindia.com (Ministry of Urban development—Migration details).

http://www.slumdetailsofindia.com (primary census abstract for slum 2011, Office of the Registrar General & Census Commissioner, New Delhi India).

ifmo. (2013). *Megacity Mobility Culture—How Cities Move on in a Diverse World*. Institute for Mobility Research, Munich, Germany.

INCCA. (2007). Ministry of Environment and Forest, Government of India.

Innes, J.E. and Booher, D.E. (2000). *Public Participation in Planning: New Strategies for the 21st Century*. University of California at Berkley, Institute of Urban and Regional Development. University of California, USA.

International Transport Forum. (2011). *Road Safety Annual Report*. http://internationaltransportforum.org/irtadpublic/pdf/11IrtadReport.pdf (Last accessed 25.05.2013).

Irwin, N.A. and von Cube, H.G. (1962). Capacity restraint in multitravel mode assignment programs. *Bulletin of Highway Research Board*, 347, 258–289.

Kittleson Associates, Urbitran Associates, LKC Consulting Services, MORPACE International, Queensland University of Technology and Y. Nakanishi. (2003). *A Guidebook for Developing a Transit Performance Measurement Systems, Transit Cooperative Research Program Report 88*. Transportation Research Board, Washington DC.

Litman, T. (2011). Transportation Elasticities—How Prices and Other Factors Affect Travel Behavior, Report, Victoria Transport Policy Institute.

Litman, T. (2012). Toward More Comprehensive and Multi-modal Transport Evaluation, Research paper, Victoria Transport Policy Institute.

Lohia, S.K. (2008). Urban Transport in India., Proc., Indo-US conference on Mass Transit Travel Behavior Research 08 (MTTBR-08), IIT Guwahati, India.

Mackay, M. (2004). A Comparison of the Sample Survey and Open House Methods of Public Consultation for Transportation Issues in the City of Edmonton. Unpublished MSc Thesis, Department of Civil Engineering, University of Calgary.

Manuel, V.S., Perez, D.V., Ramon, M.R. and Tomas, S. (2009). Public transport funding policy in Madrid: Is there room for improvement? *Transport Reviews*, 29(2), 261–278.

Mayo, J. and Tabaoda, E.B. (2020). Ranking factors affecting public transport mode choice of commuters in an urban city of a developing country using analytic hierarchy process: The case of Metro Cebu, Philippines. *Transportation Research Interdisciplinary Perspectives*, 4, 100078.

McKinsey & Company. (2008). *India's Urban Awakening: Building Inclusive Cities, Sustaining Economic Growth.* McKinsey Global Institute, McKinsey & Company

Mills, G. (1991). Commercial funding of transport infrastructure: Lessons from some Australian cases. *Journal of Transport Economics and Policy*, 25(3), 279–298.

Ministry of Housing and Urban Affairs, Government of India. Available at: https://mohua.gov.in

Ministry of Road Transport & Highways. (2008–2011). *Road Accidents in India.* Government of India. Ministry of Road Transport and Highways, Transport Research Wing, New Delhi.

Ministry of Urban Development. (2008). UMTC (2012); RITES (2012).

Mitchell, C. and Town, S. (1976). *Accessibility of various Cosial Groups to Differing Activities*. Transport and Road Research Laboratory, Crowthorne, Berks.

MMPG. (1997). Techno-economic feasibility study of construction of Mumbai metro, Progress Report Submitted by MMPG to Government of Maharashtra, India.

MoHUA. (2006). *National Urban Transport Policy*. Ministry of Housing and Urban Affairs, Government of India

MORTH. (1999, 2000, 2003). *Handbook on Transport Statistics in India*. Transport Research Office, Ministry of Road Transport and Highways, Delhi, India.

MOUD. (1998). Traffic and Transportation Policies and Strategies in Urban Areas in India, Final Report. Ministry of Urban Development, Government of India, New Delhi.

MOUD. (2008). *Study on Traffic and Transportation Policies and Strategies in Urban Areas in India*. Government of India.

Mumbai Metropolitan Region Development Authority (MMRDA). (2000). Selection of a mass rapid transit system for Mumbai, Final Report for Greater Mumbai: The Andheri-Ghatkopar Corridor, Mumbai, India.

Mumpower, J.L. (2001). Selecting and evaluating tools and methods for public participation. *International Journal of Technology, Policy, and Management*, 1(1), 66–77.

Naganna, N. and Ramanayya, T.V. (1985). *A Critical Appraisal of Road Accessibility in Rural Areas*, Presented at National Seminar on Roads and Road Transport Rural Areas, Central Road Research Institute, New Delhi.

National Rural Infrastructure Development Agency. (2023). *Annual Report for the Year 2022–23*. Ministry of Rural Development, Government of India.

Ortuzar, J. de D. and Willumsen, L.G. (1999). *Modelling Transport*. John Wiley and Sons.

Papacostas, C.S. and Prevedouras, P.D. (2002). *Transportation Engineering and Planning.* Prentice Hall of India, New Delhi.

Parasuraman, A., Zeithaml, V.A. and Berry, L.L. (1985). A conceptual model of service quality and its implications for future research. *Journal of Marketing*, 49, 41–50.

Planning Commission. (2013). *Recommendations of Working Group on Urban Transport for 12th Five Year Plan.* http://planningcommission.nic.in/aboutus/committee/wrkgrp12/hud/wg%20urban%20Transport.pdf (Last accessed 06.10.2013).

Planning for Non-motorized Transport in Cities—DIMTS-UITP Symposium, 2010; DDA (2007); Central Road Research Institute, Delhi; Bruhat Bengaluru Mahanagara Pa—like; Mobility Indicators, DULT, 2010–11; UMTC (2012); Indore Municipal Corporation; RITES (2012); Wilbur Smith Associates (2008); Guwahati Metropolitan Development Authority.

Population Projections for Indian States. Available at: https://www.india.gov.in

Pucher, J., Markstedt, A. and Hirschman, I. (1983). Impacts of subsidies on the costs of urban public transport. *Journal of Transport Economics and Policy*, 17(2), 155–176.

Pucher, P., Korattyswaropam, N., Mittal, N. and Ittyerah, N. (2005). Urban transport crisis in India. *Transport Policy*, 12, 185–198.

Ramanayya, T.V. (2005a). Strategies for Improving profitability of KSRTC, Project Report, IIM Bangalore, India

Ramanayya, T.V. (2005b). Holistic Organisational Transformation Strategies (HOTS) for APSRTC, Project Report, IIM Bangalore, India

Ramanayya, T.V. and Anantharamaiah, K.M. (1989). Impact of Facility on Economic Development in Rural Areas. In Reddy, K.V. (ed.), *Transportation Systems.* Tata McGraw Hill Publishing Company, Delhi.

Ramanayya, T.V. and Anantharamaiah, K.M. (1993). *Transportation Characteristics in Rural Areas of Southern India*, Proceedings of Third National Conference on Transportation Systems Study—Analysis and Policy, NCOTSS-1993, Andhra University, Visakhapatnam, December 1993.

Ramanayya, T.V. and Anantharamaiah, K.M. (2002). *Changes in Rural Transport Scenario in South India (1978–1999)*, Presented in International workshop Providing Road Connectivity The PMGSY Approach, A vision to transform Rural India Bhubaneswar, Orissa, February 2002.

Ramanayya, T.V. and Anantharamaiah, K.M. (2003). *Database on Rural Travel Changes in Three Decades (A Case Study of Southern India)*, Presented at National Seminar on Integrated Development of Rural and Arterial Road Network for Socio—Economic Growth, New Delhi, December 5 & 6, 2003.

Rastogi, R. (2007). *Recent Practices in Transportation Planning and Traffic Engineering*, Lecture Notes, MDONER Sponsored Short Term Course, IIT Guwahati, January 15th to 19th, 2007.

Reddy, B. and Balachandra, P. (2012). Urban mobility: A comparative analysis of megacities of India. *Transport Policy*, 21, 152–164.

Reeven, P. Van. (2008). Subsidisation and urban public transport and Mohring effect. *Journal of Transport Economics and Policy*, 42(2), 349–359.

Rural Infrastructure in Entire South India (9 States of India), Sponsored by Ministry of Transport, Government of India, 1996–98, unpublished, IIM Bangalore.

Rural Transportation Studies in Nine States of India funded by the Ministry of Surface Transport, Government of India, 1986–89, unpublished, IIM Bangalore.

Serebrisky, T. (2009). Affordability and subsidies in public urban transport: What do we mean, what can be done? *Transport Reviews*, 29(6), 715–739.

Schafer. (1998). The global demand for motorized mobility. *Transportation Research Part A*, 32(6), 455–477.

Shepherd, A. and Bowler, C. (1997) Beyond the requirements: improving public participation in environmental impact assessments. *Journal of Environmental Planning and Management*, 40(6), 725–738.

Singh, S.K. (2005). Review of urban transportation in India. *Journal of Public Transportation*. 8(1), 79–96.

Stokes, G. and Cullinane, S. (1998). *Rural Transport Policy*. Pergamon, Netherlands.

Sulek, J.M. and Lind, M.R. (2000). A systems model for evaluating transit performance. *Journal of Public Transportation*, 3(1), 29–47.

Tolley, R. and Turton, B. (1995). *Transport Systems, Policy and Planning*. Longeman Scientific and Technical.

Transport Advisory Service. (1997). *Bus Services for Rural Communities—An audit of Villages in England*. TAS, Preston.

Tsukada, S. and Kuranami, C. (1994). *Value Capture: The Japanese Experience*. Financing Transport Infrastructure, PTRC, London, January 1994.

Ubbels, B. and Nijkamp, P. (2002). Unconventional funding of urban public transport. *Transportation Research Part D*, 7(5), 317–329.

Ubbels, B., Nijkamp, P., Verhoef, E., Potter, S. and Enoch, M. (2001). Alternative ways of funding public transport: A case study assessment. *European Journal of Transport and Infrastructure Research*, 1(1), 73–89.

Vaca, E. and Kuzmyak, R. (2005). Parking Pricing and Fees, Chapter 13, TCRP Report 95, Transit Cooperative Research Program, Transportation Research Board, Federal Transit Administration.

Verma, A. (2010). *Integrated Public Transportation System Planning and Modelling*. VDM Publishing House Ltd., Mauritius.

Verma, A. and Dhingra, S.L. (2000). *A Review of the State of Art of Metropolitan Transit Alternatives*, All India Seminar on Urban Mass Transit Systems, The Institution of Engineers (I), Mumbai, India, 262–270.

Verma, A. and Dhingra, S.L. (2001). Suitability of alternative systems for urban mass transport for Indian cities. *Trasporti Europei, Quarterly Journal of Transport Law, Economics and Engineering*, Trieste, Italy, Anno, VII(18), 4–15.

Verma, A. and Dhingra, S.L. (2003). Urban rail transport corridor identification. *GIS Development, The Asian GIS Monthly*, India, 7(5), 39–42.

Verma, A. and Dhingra, S.L. (2005). Optimal urban rail transit corridor identification within integrated framework using geographical information systems. *Journal of Urban Planning and Development, A.S.C.E.*, 131(2), 98–111.

Verma, A. and Dhingra, S.L. (2006). Developing integrated schedules for urban rail and feeder bus operation. *Journal of Urban Planning and Development, A.S.C.E.*, 132(3), 138–146.

Verma, A. and Hemanthini, A.R. (2021). Church Street First—Impact Assessment of Pedestrianizing an Urban Street in terms of Quality of Life, Project Report, IISc Sustainable Transportation Lab (IST Lab), Indian Institute of Science, Bangalore, India.

Verma, A., Maneesha, B., Nitwal, R.S. and Hemanthini, A.R. (2024). Scenario Evaluation for the Proposed Sub-Urban Rail Network and Metro Rail Network for Bangalore Metropolitan Region, Project Report, IISc Sustainable Transportation Lab (IST Lab), Indian Institute of Science, Bangalore, India.

Verma, A., Vajjarappu, H., Rodeja, N. and Hemanthini, A.R. (2018). Mitigation policy bundle evaluation for Bangalore, Project Report, IISc Sustainable Transportation Lab (IST Lab), Indian Institute of Science, Bangalore, India.

Verma, M., Verma, A., Ajith, P. and Sneha, S. (2013). *Urban Bus Transport Service Quality and Sustainable Development: Understanding the Gaps*, 13th World Conference on Transport Research (WCTR), Rio de Janeiro, Brazil, July 15 to 18, 2013.

Vuchic, V.R. (2005). *Urban Transit Operation, Planning, and Economics*. John Wiley & Sons, Inc., Hoboken, New Jersey.

Wetzel, D. (2006). Facing the environmental challenge: Innovative methods of financing public transportation. *World Transport Policy & Practice*, 12(1), 40–46.

White, P. (2002). *Public Transport: Its Planning, Management, and Operation*. Spon Press, London.

WHO. (2011). *Global Health Observatory Data Repository—Road Safety: National Legislation*. http://apps.who.int/gho/data/node.main.A999?lang=en (Last accessed 06.05.2013).

World Bank. (2011). http://data.worldbank.org/topic (Last accessed 06.03.2013).

Index

Note: Page numbers in *italics* indicate figures and page numbers in **bold** indicate tables on the corresponding page.

accessibility *199–200*; analytical measuring formulas (journey time/generalized cost) 197; defined 195–196; environmental/income relationships 199; policy formulation drawbacks 195; prerequisites for economic development 189; in rural transport planning 193, 203–205
additional travel *see* induced travel demand
aggregate model 97
Ahmadabad: bicycle lane in *152*; median bus lane in *147*; population growth and ranking trends **9**; requirement of buses **226**
Ahmedabad Metropolitan Transport Service (AMTS) 218, 219; cost elements for **220**, **222**; profit and loss **222**; urban undertaking for staff positioning **223**
airport taxi 21
air quality index **232**
alighting counts 157
all-weather roads (AWRs) 189, 196, 202, 208–209
Andhra Pradesh: demand for intradistrict and rural services in 205; investment in infrastructure 238; private transport operators in 160–161; World Bank assistance for rural roads 209
Andhra Pradesh State Road Transport Corporation (APSRTC) 160–161, 205, 207, 225, 227–228, 230
articulated bus 28–29, *29*; *see also* buses
automated guided transit (AGT) **25**, 38–39, *39*, 45, **46**, **47**, 48, 129
automobiles 18–19, *19*, 100, 107; non-automobile modes 115; on-street parking regulation 139; personal mobility 126; pollution by exhaust emissions 232; private 17, 110; taxation 235, 240; technological advancement 123
autorickshaws 21, *21*

Bangalore (Bengaluru) 44, 66; Bengaluru Development Authority (BDA) 66; Bengaluru Metropolitan Land Transport Authority (BMLTA) 66–67; Bengaluru Metropolitan Land Transport Authority Bill 66; Bengaluru Metro Rail Corporation Limited (BMRCL) 66, 111; Bengaluru Suburban Rail Project (BSRP) 110; BMTC (*see* Bangalore Metropolitan Transport Corporation); Bruhat Bengaluru Mahanagara Palike (BBMP) 66; City Mobility Investment Program 67; Directorate of Urban Land Transport (DULT) 66; metropolitan region 215–217; network/land-use planning mapping 81–82, *82*, *83*; peak hour traffic loading on transport network *105*; regulation of land transport 66–67; transportation in 58–59; tunnel road and double-decker corridors 110–113, *111–113*; urban passenger transport 230–233
Bangalore Metropolitan Transport Corporation (BMTC) 50, 57, 218–219; assurance dimension 56; bus service fleet size2 014–24 **62**; bus service quality 57; bus system performance 60, **60**; cost elements for **220**, **221**; financial performance 234–235; as means of public transport 58–59, 231; Namma BMTC app 60; operational statistics 61, *61–62*; profit and loss of **222**; reliability 55–56; requirement of buses **226**; tangibility 56; urban undertaking for staff positioning **223**
basic mobility 214–215; *see also* mobility
battery-operated buses 31–32, *32*; *see also* buses
beneficiary paying development taxes/funds 75
Bengaluru Suburban Rail Project (BSRP) 110
BEST *see* Bombay Public Transport
bicycles 17–18, *18*, 62–63, 68–69, 115, 123, 151, 216–217; lane in Ahmadabad *152*
binary logit model 98–100
bi-proportional algorithm 92
BMTC *see* Bangalore Metropolitan Transport Corporation
Bombay Public Transport (BEST) 218–219; cost elements for **220–221**; profit and loss **221**; requirement of buses **226**; urban undertaking for staff positioning **222**
Buchan, K. 196
buses 47, *145*, 145–147; capacity calculations **16**; compared with LRT **131**, 131–132; environmental benefits over private modes 199–201; market segmentation metrics 195; for metropolitan cities

226; norms for bus fleet **225**; percent of public sector buses **158**; perceptions and expectations of respondents **54**; required for Indian metropolitan cities2 021 **226**; revenue and profitability 230; types used by respondents **53**; urban mobility 58; *see also specific types*
bus fleet: intra and rural operations **206**; norms based on population **225**; owned by BMTC 59; upgrading 58
bus preemption of traffic signals 147–148
bus rapid transit (BRT) 39, 129, **131**, 132; characteristics/mode **132**; design goals and technology **46–47**; urban public transportation **25**
bus streets 147, *148*

capacity-restrained traffic assignment 106
capital 239–240
carpooling 150–151, *151*
carpool lanes 145–146
carriage *see* horse-pulled carriage
case studies: Bengaluru Metropolitan Region 215–217; public mass transport (PMT) 50–57; transport infrastructure, rural India 207–211, **210–212**; tunnel road and double-decker roads (Bengaluru) 110–113, *111–113*
category analysis *see* cross-classification
central area transit terminals 155
channelization of traffic 141–143, *143*
Chennai: Chennai Unified Metropolitan Transport Authority (CUMTA) 67–70, *70*; Integrated Urban Development Platform (IUDP) 68; Urban Mobility Resilience 68; Urban Project Planning and Management (UPPM) 68
City Development Index (CDI) 133
community-wide benefits 115
commuter and intercity rail 10, 11, 16, 18, 68–69, 135
commuter park-and-ride *154*
comparative analysis 44–45, **45–47**
Comprehensive Mobility Plan (CMP) 231
contra flow bus lanes *146*, 146–147
control, defined 25
cost–benefit **212**
costing, of PMT services 218–219
cross-classification 88, **88–89**
cycle rickshaw **16**, 21, 22, 24, *24*
cycling 17–18, *18*, 62–63, 68–69, 115, 123, 151, 216–217; lane in Ahmadabad *152*

Daly, A. 195
debt burden 240
Delhi Transport Corporation (DTC) 218–219; cost elements for **220–221**; profit and loss **222**; requirement of buses **226**; urban undertaking for staff positioning **223**
demand: short, medium, and long-term 113–116; travel estimation 203–205; urban travel 78–113; *see also* need
design goals **46–47**
Dhingra, S.L. 36, 50
disaggregate model 88, 97
distributions of points 128, *128*
disutility functions 98
double-decker bus 28, *28*, **32**; corridors 111, *112*
doubly constrained growth factors 91–92
doubly constrained matrix expansion problem **93**
doubly constrained model 95
DTC *see* Delhi Transport Corporation

economics, PMT 214–233; cost of services 218–219; energy efficiency 214–217; manpower in operations and management 219–224; social, economic, and environmental benefits 224–233
Edmonton 125; Key Choices Survey 126–127
electric vehicles (EVs) 134, 232
energy efficiency 214–217
environmental benefits 224–233
EVA methodology 227–228, **229**, 230
existing transportation network 119
express bus 29, *30*
expressway ramp metering 138

fair-weather roads (FWR) 208
fare integration 70, *70*
fare structures 72–73, *73*, 75, 161, 236; trade-offs among alternative **74**
financial performance 234–238
first- and last-mile connectivity 124–132
flat fare *73*, 73–74, **74**
fleet augmentation 241
fleet requirements, in urban centers 224, **225**
fleet strength 41, 42, *42*, 160, 166, 205, 224, 238
flow bus lane 146
free/all-or-nothing assignment 103
free/multipath traffic assignment 103–104, *105*
funding 71–76, 239, 242–243; innovative options 75–76; securing 115; traditional options 71–75; *see also* public mass transport

general taxes 75
generated traffic *see* induced travel demand
geographical constraints 119
globalization 10, 133
Gould, P. 195
graduated fare 73, 75

gravity distribution model 94–96; calibration of 95–96
greenhouse gases (GHGs) 1, 214
green transport 17
gross domestic product (GDP) 1, 5, 10, 49, 216; contribution by rural and urban areas **10**, *11*; values of India from 1951 to 2010 *11*
group rapid transit (GRT) 38, 48
growth-factor methods 89–94, **91**, **93**; advantages and limitations of 93–94
guidance, defined 25
guided buses 30–31, *31*

hackney carriages 20
hand-pulled rickshaws *21*, 21–22
Harman, L. 47–48
headway 26, **33**, 41
heavy and light rail 48
high-occupancy and non-vehicular travel 150, *152*; carpooling 150–151, *151*; human-powered modes 151–153, *152*; ridesharing 150–151, *151*
high occupancy vehicle (HOV) 45
high-occupancy vehicles 144; bus bays *145*, 145–146; bus preemption of traffic signals 147–148; bus stop relocation *145*; bus streets 147, *148*; carpool lanes 145–146; contra flow bus lanes *146*, 146–147; flow bus lane 146; median bus lanes 147, *147*; toll policies 148
high-performance transit modes 129–132, *130*, **131**
Hillman, M. 195
home-based (HB) trip 85; nonwork trip rates 88, **88**
horse-pulled carriage 22, *23*
human-powered modes 22, 24, 151–153, *152*
hypothetical network *104*

IIMB *see* Indian Institute of Management Bangalore
Indian Institute of Management Bangalore (IIMB) 164, 205, 207–208, 224
induced travel demand 108–109
information integration 69, *70*
infrastructure *see* transport infrastructure
institutional integration 64–65
integrated public transport network (IPTN) 201
integrated transport infrastructure 43–44, **44**
intermediate public transport *see* paratransit
intermodal terminals 68–69
intermodal trip 43, *43*
intradistrict services 205
investment in infrastructure 238
Irwin, N. A. 104

journey 85; characteristics of 97; per person per year and purpose **196**

Karnataka State Pollution Control Board (KSPCB) 231–232
Karnataka State Road Transport Corporation (KSRTC) 207
Kerala State Road Transport Corporation (KSRTC) 161, 162, 205

land use 12, 15, 119; categories **87**; characteristics 81; importance in trip generation 84, 85; network city interactions 119; non-residential 85; planning 67; proposed map of Bengaluru, for 2031 *82*; qualified 65; transportation network and 63
latent demand *see* potential users
light rail transit (LRT) 18, **25**, 26, 33, *34*; capacity of **34**; characteristics **40**, **132**; compared with buses **131**, 131–132; compared with RRT 130, **131**; design goals and technology **46–47**; recommended MTS for cities **37**
limousines 21
line capacity 30, 41; capacity of LRT **34**; capacity of RGR **36**; capacity of RRT **35**; defined 41
Litman, T. 214
load count 157
long-range urban transportation infrastructure *80*
long-term planning 78, 114–116
loss-making routes 166, **227**

Madhya Pradesh State Road Transport Corporation (MPSRTC) 160
magnetic levitation 37–38
manpower: costs 237–238; operations and management 219–224, **222–223**
map: Indian states *2*; land-use 81, *82–83*
mass rapid transit system (MRTS) 33–37, 110, 113
mass transit *see* public transport
maximum load sections (MLSs) 157
median bus lanes 147, *147*
medium-term planning 114
mergers 65
metro *see* rail rapid transit (RRT)
metropolitan planning organization (MPO) 78
migration to urban areas 2–3, 8–9, **9**
million-plus cities **9**, 13
minibus 20, 27, *27*, **32**, 125
Mitchell, C. 196
MMR, desire line diagram of *90*
mobility 195, 214, 215; *see also specific types*
modal split 85, 96–103; aggregate and disaggregate models 97; binary logit

model 98–100; factors influencing choice of 96–97; nested logit model *100–101*, 100–103; trip-end modal split model 97; trip-interchange modal split model 97; utility and disutility functions 98
modes of transport *see* transport modality
monorails 38, *38*
motorcycles **16**, 17–18
motor vehicles (MVs) *see* automobiles
multinomial logit (MNL) *100*, 100–103
multiple linear regression technique 85–87

national guiding framework (NGF) 201
need: for funding 242–243; for network planning 193–203; *see also* demand
nested logit model *100–101*, 100–103
network city interactions 119
network integration 69–70, *70*
network operating efficiency 118–119
network planning 193–203; dedicated public transport for students 194; market segmentation 195; mobility and accessibility 195–199, **196**, **197**, *198–199*; public transport network development 199–201, **201**; India, rural transport network development in 202–203; South Africa, rural public transport network in 201–202; transport for schoolchildren and college students 194
non-automobile modes 115
non-home-based (NHB) trip 85

occasional users 115
off-road transport 15, 157
off street loading 144
off-street terminal 155
one-way street 138–139
on street parking, removal of 140–141
operating efficiency 118–119
optimum TU capacity 119–121, *121*
organization 133–155; high-occupancy and non-vehicular travel modes 150–153; parking management 153–155; preferential treatment of high-occupancy vehicles 144–148; reduced peak period travel 148–150; role in solving issues of urban transport 133–134; state-level for resource use 242–243; of surveys 156; transit and paratransit service 155; transportation system management (TSM) 134–136; vehicular flow (*see* vehicular flow, improved)
Orissa State Road Transport Corporation (OSRTC) 160
outlying terminal 155
ox wagon 23, *23*

paratransit 19–20, **25**, 26, 136, 144, 155
park-and-ride facilities 154
parking 136, 139, 141, 153–155
passenger attraction 117–118
passenger flow 41
passenger volume 41, **132**; and load count 157
peak-period flow 148–149; truck restrictions 150; work rescheduling 149–150
people mover systems (PMS) 38
performance-cost package (PCP) 129, *130*
personal rapid transit (PRT) 38
personal transportation 17–19
persons per household 88, **89**
physical integration 68–69
planning and management 133–168; assessment of user satisfaction 167; data collection, retrieval, and analysis 156–157; global trend in government policies 167–168; organization involved for (*see* organization); private sector role in 157–166; *see also individual entries*
planning PMT in rural areas 189–213, **197**; network planning in 193–203; PMGSY 189–192, **192**, *193*; rural transport impact on economic development 189; stakeholder consultation and preferences for optimizing 205–207; transport infrastructure 207–211, **210–212**; travel demand 203–205
planning PMT in urban areas 78–132; choice of transportation mode 122–123; concept of LOS in 123–124; process in organization *79*; routing and scheduling 116–121; stakeholder consultation and preferences 124–132; transport infrastructure 113–116; urban travel demand 78–113
PMGSY *see* Pradhan Mantri Gram Sadak Yojana
polluter paying carbon taxes/funds 76
population growth 133; in developed and developing countries *6*; historical growth of **5**; since 1951 **4**
potential users 115
Pradhan Mantri Gram Sadak Yojana (PMGSY) 189–192, *193*; addressing rural road connectivity 202; road accessibility 196; road works *193*; yearly sanction details of **192**
private hire vehicles 21

private sector, in PTM: competitive advantages of STUs 158–159, 164; controlling domination of 162; cost structure/economics **163**, 163–164; easy entry/easy exit 161; factors influencing 161–164; faster pay back 162; in India 157, **158**; low investment/low technology 161–162; operators 159–161; policy issues faced by 164–165; privatization 165–166; quality parameters 164; revenue model 162–163
propulsion, defined 25
public bicycle sharing 18
public mass transport (PMT) 49–77; case study 50–57; in developed and developing nations 49–50; economics and efficiency parameters 214–233; first-mile and last-mile connectivity and loss 62–70; funding (*see* funding); planning and operational strategies 57–62; shortcomings 59–60; state-level resource management 234–238; *see also* planning and management
public transport 26–39, 134; classification of 26–39; coverage 42, **43**; defined 26; environmental benefits of 230–233; functions of 47–48; network development 199–201, **201**; operational and infrastructural taxes/fund 76; rapid transit 33–37, **37**; semi-rapid transit 33; special transit 37–39; status in urban areas 42, *43*; street transit 27–31, **32–33**; systems 13, **37**, 45, 62, 72, 125, 214; *see also individual entries*

rail-based transit, characteristics of **40**
rail rapid transit (RRT) 34, 36, 129; capacity of **35**; compared with LRT 130, **131**; design goals and technology **46**; mass transit systems **37**; *see also* light rail transit (LRT)
Rajasthan State Road Transport Corporation (RSRTC) 159–160
rapid transit 26, 33–37, **37**
Rastogi, R. 17, 18
regional rail transit (RGR) 36, **36–37**
regular bus (RB) 27–28, *28*, 131, **131**
research methodology 111
reversible lanes 141, *142*
rickshaw: autorickshaws 21, *21*; cycle rickshaw **16**, 21, 22, 24, *24*; hand-pulled rickshaws *21*, 21–22
ridership 61, *61*, 110–112, *113*, 241–242
ridesharing 150–151, *151*
rights-of-way (ROWs) category 15, 24–25, **25**, 33, 129, *130*
Road Transport Corporations Act 50
rolling stock 238
routing and scheduling 69; categories 117; planning PMT in urban areas 116–121
rubber-tired rapid transit (RTR) 34, *35*
rural settlements: global spread of 202; population in 3, *3*, **4**, 10, 189, **190**, 205; urbanization 3
rural transport: development through PMGSY 189–192, **192**, *193*; impact on economic development 189; network development in India 202–203; policy requirements *198*; problem *198*; public 193, **201**, 201–202; *see also* planning PMT in rural areas

scheduling *see* routing and scheduling
sectional fare **74**, 74–75
semi-rapid buses (SRBs) 33
semi-rapid transit **16**, 33
service quality: attributes **207**; gaps in 53–55
SERVQUAL framework 50–57; aim of 51; assurance dimension 56; demographic profile of respondents 52, **52–53**; empathy dimension 56; expectations and perceptions **54**, *55*; reliability dimension 55–56; responsiveness dimension 56; sample profile 52, **52–53**; sampling 51; service quality gaps 53–55; tangible dimension 56; *t*-test analysis 56–57, **57**; variable measurement 51
settlement size **211**
short-term planning 114
signalized intersections 137
signalizing expressway ramp 138
singly constrained growth factor 91, **91**
singly constrained model 95
slum population 8, *191*
social benefits 224–233
South Africa, rural public transport network in 201–202
special transit 37–39
speed-and-delay survey 156
stakeholders 124–132; consultation and preferences 124–128, 205–207
state revenue 242–243
state transport undertakings (STUs) 42; capital contribution to 239; capital structure and investments 239; competitive advantages of 158–159, 164, **204**; financial performance of 234–237;

fleet strength from 2010–11 to 2019–20 *42*; forms of 206; manpower costs in 237–238; personnel costs in **237**
strategic objectives 115
strategic plan *see* long-term planning
streetcar (SCR) 29–30, *31*
street transit 27–31; capacity of **33**; system parameters for **32**
students' concessions 240
support, defined 25
surface transit *see* street transit
synthetic model *see* gravity distribution model
system parameters, for street transit **32**
system technology 25, 129

tariff association 64
taxi-buses 21
taxicab *20*, 20–21
technology attributes **46–47**
Thane, peak hour traffic loading *107*
toll policies 148
topography and environment 119
Town, S. 196
tramway 29–30, *31*
transit, defined 26
transit communities 64
transit federations 65
transit integration 64–65
transit mode, characteristics 24–26
transit modes, feasibility and comparison: bus rapid transit (BRT) 129; comparison of BRT and LRT 132, **132**; high-performance transit service 129; light rail compared to buses 131, **131**; performance-cost package (PCP) 129, *130*; ROW categories 129; RRT compared LRT 130
transit network design 116–117
transit shelters [illegible]
transit stop relocation 144
transit system: capacity, work and utilization 41; operational factors in 39–42
transit terminals 155
transit unit (TU) 41; maximum load 157; optimum capacity 119–121, *121*
transit way *see* rights-of-way (ROWs) category
transportation improvement program (TIP) 134
transportation mode: choice of 122–123; classification of **25**
transportation system management (TSM) 78, 134; advantages 135; classification 135–136
transport infrastructure: basic steps in planning *80*; characteristics of 97; comprehensive urban planning 79, *80*, 115, 134; development of rural India 207–211, **210–212**; integrated transport approach 43–44, 62, 64, 107; need for 43–44; short, medium, and long-term demand 113–116; successful integration of 64–65
transport modality 15
transport model, classical: induced travel demand 108–109; modal split 85, 96–103; travel demand modeling 106–108; trip assignment 85, 103–106; trip distribution 85, 89–96; trip generation 84, 85–89
transport services: classification of 204, **204**; concept of LOS in 123; cost elements **220**; easy entry and exit 161; funding of 26; subsidies 26, 71; sustainable 49
travel demand model (TDM) 106–108, **108**, 110, 203–205
travel time 41–42, 95–96, 122–124
trip assignment 85, 103–106; capacity-restrained traffic assignment 106; free/all-or-nothing assignment 103; free/multipath traffic assignment 103–104, *105*
trip attraction 85, 87, 92, 96
trip distribution 85, 89–96; gravity distribution model 94–96; growth-factor methods 89–94, **91**, **93**
trip-end modal split model 97
trip generation 84, 85–89; cross-classification (or category analysis) 88, **88–89**; multiple linear regression technique 85–87; trip-rate analysis 87, **87**
trip-interchange modal split model 97
trip maker 85, 96
trip production 85, 87, 88
trip-rate analysis 87, **87**
trolley bus (TB) 29, *30*
truck restrictions 150
turn restrictions 110, 116

uniform growth factor 89, **90**
universal service obligations 240–241
urban areas 42–43; challenges 13; classification 6; contribution of GDP **10**; criteria for defining 6; defined 3–5; migration to 2–3, 8–9, **9**; population *3*, **4**, 5, **5**, 12, 216; share of public transport trips 17; *see also* planning PMT in urban areas
urbanization 3–8, 10, 50, 62, 114, 133
urban passenger transportation: classification by carrier type **16**; classification by type of usage **16**
urban transport **220**; challenges 13; classification 15–24, **16**; in developing countries 10–13; efficient 214; infrastructure 79;

long-range infrastructure *80*; mobility 68, 123–124; *see also* planning PMT in urban areas
utilization 41, 98; channelization of available space 142; fleet utilization 42

vehicle capacity **34**, **35**, **36**, 41, 121
vehicle miles travel (VMT) 109
vehicular flow, improved 136–137; channelization of traffic 141–143, *143*; off street loading 144; one-way street 138–139; on-street parking, removal of 139–141; reversible lanes 141, *142*; signalized intersections 137; signalizing expressway ramp 138; transit stop relocation 144; turn restrictions 143
Verma, A. 36, 50, 116, 215
von Cube, H. G. 104
Vuchic, V. R. 119

waterborne transport 38
work rescheduling 149–150

zonal fare *73*, 73–74, **74**

For Product Safety Concerns and Information please contact our EU
representative GPSR@taylorandfrancis.com
Taylor & Francis Verlag GmbH, Kaufingerstraße 24, 80331 München, Germany

www.ingramcontent.com/pod-product-compliance
Ingram Content Group UK Ltd.
Pitfield, Milton Keynes, MK11 3LW, UK
UKHW022223190926
475702UK00003B/480
* 9 7 8 1 0 3 2 6 6 1 9 9 5 *